Markus Beck

The Renewable Energy Directive and the challenges for the Biodiesel Industry

Insigths into a dynamic and opaque industry

Anchor Academic
Publishing

Beck, Markus: The Renewable Energy Directive and the challenges for the Biodiesel Industry: Insigths into a dynamic and opaque industry. Hamburg, Anchor Academic Publishing 2013

Buch-ISBN: 978-3-95489-065-1
PDF-eBook-ISBN: 978-3-95489-565-6
Druck/Herstellung: Anchor Academic Publishing, Hamburg, 2013

Bibliografische Information der Deutschen Nationalbibliothek:
Die Deutsche Nationalbibliothek verzeichnet diese Publikation in der Deutschen Nationalbibliografie; detaillierte bibliografische Daten sind im Internet über http://dnb.d-nb.de abrufbar.

Bibliographical Information of the German National Library:
The German National Library lists this publication in the German National Bibliography. Detailed bibliographic data can be found at: http://dnb.d-nb.de

Hermannstal 119k, 22119 Hamburg
http://www.diplomica-verlag.de, Hamburg 2013
Printed in Germany

Abstract (English)

In response to climate change and unsustainable energy consumption, the European Parliament launched a climate and energy package in 2009. This included the 20:20:20 Energy Strategy aimed at decreasing the effects of climate change, in particular by lowering greenhouse gas (GHG) emissions by 2020. This 10-year action plan proposes measures to increase the current efficiency levels of energy use and raises the share of renewable energies within the energy mix for all 27 EU Member States. As a result the increased use of biofuels, particularly in the transport sector, will be an important part of a more complex framework. Although biofuels represent only two per cent of total transport fuels used, political incentives, technology and efficiency improvements could increase this by eight per cent in Europe by 2020.

In line with the 20:20:20 Strategy, the Renewable Energy Directive (RED) was introduced to regulate the overall biofuel market, amongst others. Based on the RED's definition of sustainability, different certification schemes emerged, aimed at creating a uniform biofuel standard within Europe. At present, the overall scale and scope of environmental and social impacts associated with the biofuel supply chain are not well defined and this will be evaluated within this work. Furthermore, close ties between the biofuel supply chain and other sectors including food and crude oil industries lead to associations with issues such as deforestation, pollution and food supply shortages. Thus, this book will analyse whether RED actually supports such a low-carbon pathway or it mainly supports the local industry. As such, the RED has been critically regarded as a market protectionist mechanism for the European agricultural and biofuel industry against the comparatively more competitive biodiesel imports from Argentina and Indonesia. Finally, the EU set a tone with its Directive, encouraging international collaborations with other governments to work towards sustainable energy measures as part of a universal climate change solution.

Résumé (Français)

En réponse aux changements climatiques actuels et à la consommation d'énergies fossiles, le Parlement Européen a lancé un ensemble de mesures relatives au climat et à l'énergie. Parmi elles, l'initiative 20:20:20, qui a pour but, entre autres, de diminuer de 20% les émissions de gaz à effet de serre d'ici à 2020. Ce plan d'action sur 10 ans prévoie d'améliorer l'efficacité énergétique des sources d'énergies actuelles ainsi que d'augmenter à 20% la part des énergies renouvelables dans le bouquet énergétique des 27 membres de l'Union. Parallèlement, l'utilisation croissante des biocarburants, notamment pour les transports, fait partie intégrante d'une structure générale plus complexe. Bien que les biocarburants ne représentent que 2% des carburants utilisés par le secteur des transports, des mesures d'incitation, tant au niveau politique que technologique ainsi que des progrès en terme d'efficacité pourrait mener cette part à 8% d'ici à 2020 dans l'Union Européenne.

En accord avec l'initiative 20:20:20, la Directive Européenne 2009/28/CE sur les Energies Renouvelables (DER) a été mise en place avec pour objectif, parmi d'autres, de réguler le marché des biocarburants. Fondés sur la définition de "renouvable" énoncée dans la directive, plusieurs schémas de certification ont fait leur apparition, dans le but de créer un standard européen des biocarburants. Actuellement, l'échelle ainsi que l'étendue des impacts environnementaux et sociaux de la chaîne de production et d'acheminement des biocarburants ne sont pas encore définis, et ce travail de recherche propose de s'y essayer. En outre, de nombreux liens existent entre la chaîne des biocarburants et d'autres secteurs comme l'agro-alimentaire ou l'industrie pétrolière, ce qui provoque des associations avec des concepts problématiques tels que la déforestation, la pollution de l'environnement et la pénurie de denrées alimentaires. Ainsi, il s'agit de déterminer si la DER soutient le développement de solutions faiblement émettrices de carbones ou se contente simplement de soutenir l'industrie locale. La DER a déjà été critiquée d'être un mécanisme protectionniste déguisé pour le marché agricole Européen et l'industrie des biocarburants à l'encontre des biocarburants plus compétitifs importés d'Argentine et d'Indonésie. En définitive, l'Union Européenne a instauré un cadre général grâce à la DER encourageant les coopérations entre gouvernements à l'échelle internationale afin de travailler ensemble à des mesures en faveur des énergies renouvelables, dans le cadre de la lutte globale contre le réchauffement climatique.

Resumo (Português)

Em resposta as mudanças climáticas e ao consumo insustentável de energia, o Parlamento Europeu lançou um pacote de medidas para esses problemas. Isso inclui a Estratégia de Energia 20:20:20 visando a redução dos efeitos das mundanças climáticas particularmente através da diminuição de emissão de gases do efeito estufa até 2020. O plano de ação para 10 anos propõe medidas para aumentar o atual nível de eficiência do uso de energia e elevar o share das ações de energias renováveis dentro do mix the energia para todos os 27 estados membros da UE. Além disso, o aumento do uso de biocombustíveis, particularmente no setor de transporte será uma parte importante de uma estrutura mais complexa. Embora os biocombustíveis representem apenas dois por cento do uso total de combustiveis em transportes, incentivos políticos e melhoras de tecnologia e eficiência podem aumentar esse número para 8 por cento na UE até 2020.

Em acordo com a Estratégia 20:20:20, a Diretiva de energia renovável (RED) foi apresentada para regulamentar o mercado de biocombustível como um todo, entre outros. Baseado na definição de sustentabilidade do RED, camadas diferentes de certificaçao surgiram, com o objetivo de criar uma padronização do biocombustível dentro da Europa. No momento, em geral a escala e o efeito do impacto ambiental e social causado pela cadeia de biocombustivel nã está bem definida e isso será avaliado dentro desse trabalho scientifico. Além disso, close ties entre a cadeia de suprimento de biocombustivel e outros setores incluidon alimentos e a industria de óleo cru são associadas com problemas como desmatamente, poluição do suprimento de alimentos. Assim, se a RED realmente apoia o caminho da baixa emissao de carbono ou apoia principalmente a industria local, fica a questão. Tão como, a Red tem sido criticamente considerada como um mecanismo de mercado protecionista para a agricultura e industria de biocombustível Europeu contra uma maior competitividade de importação de biodiesel da Argentina e Indonésia. Finalmente, a Europa estabeleceu um discurso com a sua Diretiva, encorajando a colaboração internacional com outros governos para trabalharem a favor das medidas da energia sustentável como parte de uma solução universal para mudança climática.

Table of contents

List of Figures

List of abbreviations and acronyms

ACTI	Alfred Carl Toepfer International (Supply Chain Manager/Trading company)
ADM	Archer Daniel Midland
AD	Aggregated Demand
B99	Biodiesel blend with fossil diesel at a 99% biodiesel content ratio
Bio-SG	Bio-synthetic gas
BMELV	Bundesministerium für Ernährung, Landwirtschaft und Verbraucherschutz
BMU	Bundesministerium fuer Umwelt (Environmental minisry)
BtL	Biomass to Liquid
CAP	Common Agricultural Policy
CAPRI	Common Agricultural Policy Regionalized Impact model
CBOT	Chicago Board of Trade
CDM	Clean Development Mechanism
CES	Constant elasticity of substitution
CFPP	cold filter plugging point
CH4	Methane
CNFAP	Centre for National Food and Agricultural Policy
CO2	Carbon Dioxide
COCERAL	Comité du Commerce des céréales, aliments du bétail, oléagineux, huile d'olive, huiles et graisses et agrofournitures
COCO	Complete and consistent regional database of CAPRI
COMEXT	External trade statistic division of EUROSTAT
CSR	Chinese State Reserve
CSTR	Continuous Stirred Tank Reactor
CTG	Craddle-to-Grave
DDGS	Distillers Dried Grains with Solubles
Diff.	Difference
e.g.	for example
EBA	initiative Everything But Arms initiative
EBB	European Biodiesel Board
EBIO	European Bioethanol Fuel Association
EIA	UK Environmental Investigation Agency
EIA	US Energy Information Administration
EPA	Environmental Protection Agency
EPURE	European Renewable Ethanol Association (former EBIO)
EREC	European Renewable Energy Council
ESG	Environmental Services and Goods
EU 27	European Union with its 27 member states
EUROSTAT	Statistical Office of the European Communities
EX	Export
FAME	Fatty Acid Methyl Esters
FAO	Food and Agriculture Organization of the United Nations
FAPRI	Food and Agricultural Policy Research Institute

FEDIOL	European Union Oil and Proteinmeal Industry
FER	Fossil Energy Ratio
FFA	Free Fatty Acid
FOB	Free on Board
FQD	Fuel Quality Directive
FSC	Forest stewardship council
GDP	Gross domestic product
GHG	Greenhouse gas
GMO	Genetically Modified Organism
GTAP	Purdue University, Centre for Global Trade Analysis
GTIS	Global Trade Information System
ha	Hectare
HVO	Hydro treated vegetable oil.
ICE	Intercontinental Exchange
i.e.	it est
IEEP	Institute for European Environmental Policy
IFPRI	International Food Policy Research Institute
IfW	Kiel Institute for World Economy
IIASA	International Institute for Applied Systems Analysis
ILUC	indirect land use change
IM	Import
IPCC	International Panel on Climate Change
IPC	International Food and Agricultural Trade Policy Council
IPTS	Institute for Prospective Technological Studies
ISCC	International Sustainability and Carbon Certificate
k	Kilo
Kl	Klimaschutz
kmt	kilo metric tons
l	litre
LCA	Life Cycle Assessment
LCI	Life Cycle Inventory
LDC	Louis Dreyfus Commodities
LEI	Agricultural Economics Research Institute
MBS	Mass Balance System
mha	millions of hectares
MJ	Mega Joules
mmt	Million metric tons
MS	European Member States
mt	Metric ton
N2O	Nitrous Oxide
N2o	Nitrous Oxide (Laughing gas)
NBSA	Noble Brasil S.A.
NER	Net Energy Ratio
NG	Noble Group Ltd.
NPO	non-profit organization
NREL	National Renewable Energy Laboratory
NRSA	Noble Resources S.A.
OECD	Organization for Economic Co-Operation and Development
OPEC	Organization of Petroleum Exporting Countries
PFR	Plug Flow Reactor
PME	Palm oil Methyl Ester
PO	Palm oil
POLES	Prospective Outlook on Long-Term Energy Systems model
PPM	Process and Production Method
PTW	Pump-to-Wheel
RE	Renewable energy
RED	European Renewable Energy Directive (2009/28/EC)
RME	Rapeseed Methyl Ester
RSB	Round Table on Sustainable Biofuels
RSBA	Abengoa RED Bioenergy Sustainability Assurance
RSO	Rapeseed oil
RSS	Rapeseed seed
RTRS	Round Table on Responsible Soy
RTSPO	Round Table on Responsible Palm Oil
SAN	Sustainable Agriculture Network
SB	Soybean
SBM	Soybean meal
SBO	Soybean oil
SCM	Supply Chain Management
Sel.	Selected

SFO	Sunflower oil
SFS	Sunflower seed
SME	Soybean Methyl Ester
SMEs	Small and Medium sized Enterprises
SOE	State owned enterprises
toe	Ton Oil Equivalent
TRQ	Tariff rate quota
UCO	Used Cooking Oil
UK	United Kingdom
UN	United Nations
UNEP	United Nations Environmental Programme
USA	United States of America
USDA	United States Department of Agriculture
VAT	Value added tax
VETC	Volumetric exercise tax credit
WRI	World Research Institute
WTO	World Trade Organization
WTP	Well-to-pump
WTW	Well-to-wheel
WWF	World Wide Fund

1. Introduction

The twenty-first century bears new social, economical and environmental challenges for the global community. Many of these are directly linked to the world's resources, which have formed a profound basis for economic growth and development particularly in recent decades. Even though the demand for food, energy and electricity has been increasing, supply could be balanced through productivity enhancements mainly driven by technological development. This supply-demand equilibrium has become increasingly unstable and the era of cheap resources is coming to an end. As early as 1798, Thomas Malthus expressed his concerns about poverty and famine caused by the lack of food supply for an increasing world population in his famous "essay on the principle of population".[1] Nowadays, the extraction of resources or cultivation of agriculture is driven by price inflation, price volatility and environmental limitations so that a sustainable supply for the continuously rising world population cannot be maintained, and discrepancies between rich and poor on international as well as national scales will become even more dramatically recognizable.[2] There has been unprecedented progress in fighting poverty and promoting development over the last two decades, but the shift in the priorities of developing and emerging countries away from climate change, represents a dramatic threat and jeopardizes further this progress.[3] Furthermore, the apparent incompatibility between the need to reduce the impact of global warming and the resolution of the growing energy demand worldwide creates new pressure on governments to intervene politically, especially after the Stern Report[4], published in 2010, and the United Nations Intergovernmental Panel of Climate Change (IPCC) confirmed that the increase of greenhouse gas (GHG) emissions is, with a 90 per cent certainty, produced by anthropogenic activities. Among other initiatives like Clean Development Mechanisms (CDM) and Carbon Emission Trading Schemes, alternatives to nuclear and fossil energy, such as biofuels, have been promoted with subsidies

[1] Malthus, 1970.
[2] Dobbs, 2011.
[3] Steer, 2011.
[4] Stern, 2010.

and regulations in the transport sector. Carbon emissions and global warming can thus be reduced, thereby mitigating the impact of GHG.[5]

In 2011, the WWF published The Energy Report[6], which stressed the possibility of having a realistic 2050 scenario in which 100 per cent of the energy supply stems from renewable resources. Without dramatic scientific improvements and current technologies the move from fossil fuel energy towards a sustainable energy solution would not be feasible. The public sector would need to provide certain frameworks to be complemented with private sector initiatives and considerable capital-intensive investments.[7] During 2008-2010, US$ 400 bn was invested in clean energy[8] including energy efficiency and transport-related energy initiatives by the private sector.[9] Approximately 50 per cent of these investments have taken place in the developing world. Nevertheless, if the global community aims at a two degree world, that is, one in which global warming will be limited to an increase of two degrees Celsius,[10] these efforts will have to triple in the upcoming years. In addition to the private sector, more than 100 public institutions commenced with precise regulations towards renewable energies in order to keep up with the rise of the other two to three billion more people that will shift into the demand function by 2020 and an additional two billion by 2050.[11]

One of these actions is the 20:20:20 Energy Strategy initiated by the European Parliament which refers to improvements that are intended to be achieved by the year 2020: 20 per cent less overall GHG emission compared to 1990 levels; 20 per cent efficiency increase in the current energy use; and finally 20 per cent as a share of renewable energy within the energy mix used by the European Union.[12] This ambitious energy strategy is directly linked to current forecasts for energy supply and demand.

[5] European Parliament and Council, 2001 and 2003.
[6] WWF, 2011.
[7] WWF, 2011.
[8] excluding wind power investments.
[9] REN21, 2010 and UFOP, 2011
[10] Alcamo, 2009.
[11] WWF International, 2011 and United Nations Department of Economic and Social Affairs, 2006..
[12] Shane, 2012.

Global energy consumption is predicted to increase by 30 per cent by 2030 due to the growing world population and strong economic growth in developing and emerging countries. Particularly notable will be the increase of energy consumption in China and India.[13] Demand will be further driven by the rise of a considerable middle class in emerging countries such as Brazil, Colombia, South Korea, Thailand and Malaysia, and reinforced by urbanization and industrialization mainly in Asia and Africa.[14] In 2007, China exceeded the United States in terms of industry size[15], simultaneously became the largest emitter of carbon dioxide equivalent (CO2eq) in absolute terms. This demonstrates the continuously increasing pressure on the Earth's eco-system.

In light of this, through the 20:20:20 strategy, Europe aims to set a new global benchmark for sustainable energy policy.[16] Its strong encouragement of innovation and promotion of alternative energy sources also aims to reduce dependency on fossil fuels in the EU27[17], which is a particular concern given the aforementioned volatility in oil and energy prices. These alternative energy sources would, in theory, contribute to a more balanced energy security and less dependency on Middle Eastern oil and Russian gas.[18]

In line with the third objective of the 20:20:20 Energy Strategy regarding energy mix, the EU27 uses one measure to reduce carbon emissions within the transport sector. Transport has demonstrated the fastest GHG increase in Western countries including the USA, Japan and Europe.[19] It now accounts for 20 per cent of global GHG emissions and is expected to further grow by 55 per cent by 2030.[20] Thus the European parliament passed within the RED clear instructions to increase the share of biofuels up to ten per cent within this sector in order to foster cleaner energy alternatives in non-fossil fuel sources.[21] In recent years, biofuels have become a feasible alternative to oil within the transport sector and in countries like Brazil the public transport sector already runs on almost 100 per

[13] EC, 2006.
[14] IADB, 2007.
[15] That is Industrial sector that contributes to the GDP
[16] EC, 2010.
[17] EU27 refers to the European Union with its 27 Member States
[18] Mathews, 2007
[19] Fitzgerald, 2012.
[20] EC, 2006.
[21] IADB, 2007

cent biofuels successfully.[22] In contrast to alternative technologies such as hydrogen, biofuels are easily adaptable to existing infrastructure, so that flexi-fuel technology[23] has become increasingly available. Finally, Brazil and the United States – the most efficient producers as in terms of biodiesel and bioethanol respectively – have competitive prices relative to the oil industry on the global energy markets.[24]

There are different generations of biofuels and first-generation biofuels comprise among others bioethanol and biodiesel, which are made with agricultural feedstock such as corn and sugar cane as well as soybean oil (SBO), rapeseed oil (RSO), sunflower oil (SFO) and palm oil (PO). These are then processed into liquid fuels. Despite maturing and having reached a developed stage in recent years, the international biofuels trade will continue to increase heavily in the medium term. An even more elevated volume of international trade in biofuels is expected to follow rising blending requirements, which are set by politicians. These blending margins are targeted to be elevated in order to secure as well as raise demand on a continuous basis. Henceforward, production supply can keep pace with demand in domestic and international markets. However, international trade will remain a major role for countries which are not able to cover blending targets with their own production in terms of capacity or costs. Export possibilities for the most efficient producers of either biofuels or feedstock are created, for example biodiesel, SBO and PO from Brazil, Argentina and Indonesia respectively.

The European market will be of increasing importance since the EU27 on a consolidated and national level will increase their blending margins and introduce requirements, such as the nine billion Euro European Industrial Bioenergy Initiative[25] or the ethanol blending margin of E10 introduced in Germany. The European market therefore significantly affects the growing feedstock and biofuel industries in the southern hemisphere on a short- and medium-term basis.[26] This said, demand from Europe will be comparatively small with possible upcoming

[22] USDA, 2010 and 2011b.
[23] Explain this technology
[24] Dufey, 2006
[25] Energy 2020, p. 19
[26] Verdonk, 2007

large-scale imports to China and India in the long run. Other Southeast Asian countries – especially Indonesia and Malaysia, which are the leading palm oil producing economies – will cover the main vegetable oil flow into the tiger and dragon economies. By the same token, Europe's share in biofuel imports in the global market is comparative small due to enormous investments in refining capacity in the past and a protected market. European import demand will be met by South American producers, as they have comparative advantages versus local producers, including land availability, right climate and soil conditions, longer growing seasons and lower cost of production.[27]

Figure 1: Development of world biodiesel production between 2000 – 2010 (mmt)[28]

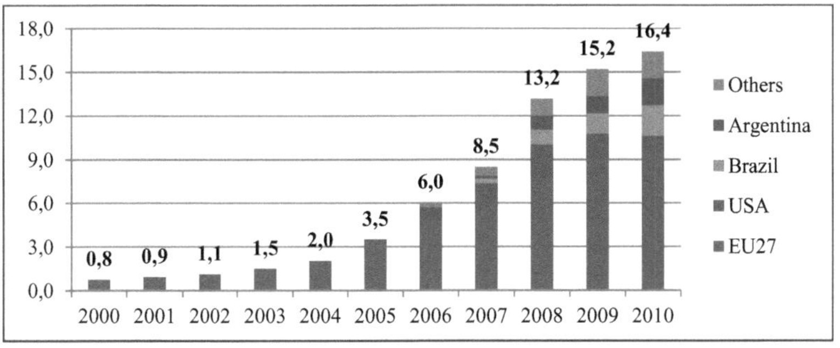

Generally over the last decade, biofuel markets have been developing in size and complexity. Through strong policy intervention in terms of regulations, subsidies and other frameworks, these have been more closely bound together in the global biofuel market, especially in the energy and agricultural sectors. In comparison to 1990, where world fuel-ethanol production amounted to about 12 million metric tons (mmt) and biodiesel production was marginal, in 2011 world fuel-ethanol production amounted already up to 80 mmt[29] and biodiesel up to 16mmt (Figure 1).[30] Ambitious future biofuel targets as stated for example by RED,[31] which set the target of 10 per cent energy from renewable sources in total European transport energy consumption in 2020, indicate that this trend will continue up to a combined global biofuel demand of 700-800mmt by 2050.[32] While Europe is

[27] IPC, 2006.
[28] Own demonstration based on European Biodiesel Board and Ufop
29 Global Renewable Fuels Alliance(2011)
[30] ENERS Energy Concept, 2010
[31] European Commission, 2009
[32] IEA, 2010.

currently the most important biodiesel producer with a share of almost 56 per cent, the production of ethanol takes place mainly in America, especially in the U.S. (approximately 58 per cent) and Brazil.[33]

In addition to positive economic and environmental effects triggered by the substitution of fossil fuels with biofuels, discussions on possible negative – or at least uncertain – indirect effects of growing biofuel markets arise. These effects are often associated with rising food prices and price volatility, which result in unsecured food supply or accessibility by lower classes, leading to hunger, malnutrition and conflicts among other issues.[34] At present the vast majority of biofuels, which are globally produced are first generation processing technologies, are based on agricultural feedstock, that is, commodities like grains and oilseeds as well as sugar crops. Hence, the biofuels generation is directly connected to commodity markets, its prices and their implicit volatility as well as depending on global demand and supply models. Furthermore, an increasing supply of biofuel by-products such as DDGS[35] on global feed markets can be seen as a result of the growing biofuel production.

Regarding feed markets the long-term impact of biofuel production is ambiguous. On the one hand, increasing prices of agricultural products used as feed components can lead to increasing feed prices and thus to shifts within the meat supply chain. On the other hand, growing supply of biofuel by-products, which can be used as feed substitutes, may cause opposite effects. The development of advanced biofuels could reduce the interconnectedness between agricultural markets and the energy demand. Considered as the second generation of biofuels technologies, their feedstock are based on non-agricultural food crops, such as cellulosic biomass including agricultural residues or algae oil.[36] As long as a marketable production of this second generation biofuels is only marginal due to an insufficient technological progress, large-scale biofuel production will continue impacting agricultural products, food markets and land usage. While looking at

[33] EBB, 2010

[34] In chapter **Fehler! Verweisquelle konnte nicht gefunden werden.**, the biofuels market and global trade flows will be presented

[35] Dried Distillers Grains with Solubles (DDGS) is a co-product, rich in cereal and yeast proteins, energy and vitamins that is produced along the ethanol distillation process and has an elevated nutrient (animal) feed value.

[36] Will be outlined in chapter 4.5

two markets that both have been of significant importance, the food and energy markets and at their corresponding prices, it becomes clear that changes to food prices are essentially local, whereas the oil price moves on a global scale. Nonetheless, a global integration of resource systems have been developed over the last decade and in particular the link between food and energy. Even though energy markets are 10 times the size of food markets, the volatility washes over from the energy market to the agriculture industry and from there to the related food market, especially affecting the grains and oilseed and meat markets. These issues concern governments, so regulations need to be taken on both, national and international levels.[37]

The consideration of biofuel trade is crucial to the evaluation of different political actions taken in the biofuels context. As biofuels can be transported at relative low costs per unit and production costs vary strongly between countries, it is probable that the relevance of biofuel trade will further increase. Besides the economic impacts of biofuel production, environmental aspects have recently dominated the political scene. The reduction of GHG emissions by increasing biofuels usage and therefore substituting the use of fossil fuels is an opportunistic move in the fight against global warming. Nonetheless, the GHG emissions of biofuels and their impact, in particular along the entire supply chain including the feedstock production and refinery process, raise critical questions. Based on those, politicians are meant to set certain frameworks and standardizations to form a smooth transition and guarantee a sustainable development from which all stakeholders can sustainably benefit.

1.1. Motivation and research objective

This work aims to assess and evaluate such an implementation of biofuels regulation, that is, the introduction of a certification scheme by the European Parliament, which has been introduced in line with the targets of the RED until 2020. As the book will analyse the EU27 regulation, the main focus will be on biodiesel and its corresponding feedstock source material as well as its effects on

[37] Nussbaum, 2011.

global biofuel and agricultural markets. In general, the commodity markets as well as the supply chain management industry are very opaque and certification schemes are only recently established. Thus, these circumstances create a very dynamic and entrepreneurial environment so that business patterns as well industry structures have not emerged yet. Current trends will be outlaid demonstrating a real-case scenario as being directly related and supported by the publicly-listed company Noble Group. Finally, the book should provide on the one side concrete industry insides and hands-on experience as well as on the other side convey a bigger climate change related picture and serve as a complementary education to the theory carried out during the lectures within the Master programs in Sustainable Development or International Supply Chain Management.

1.2. Methodological approach

The basis of this work is fundamental theoretical research of industry documents and academic works published over the last decade as well as the direct relationship with companies being active in the biodiesel and commodity trading industry, including Louis Dreyfus Commodities, ADM, Cargill and Noble Group among others. The book has been elaborated within the academic double degree program *Master of Science in Sustainable Development* at the French Univeristy *Ecole des Hautes Etudes Commerciale* (*HEC*) Paris and the *MBA em Gestão de Sustentabilidade* at the Brazilian University *Fundação Getulio Vargas* Sao Paulo during the academic years of 2010/11 and 2011/2012. Furthermore, the general macroeconomic and microeconomic knowledge supporting this work has been gained during economics, strategy and business degrees at the Spanish *Universidad Pontificia Comillas (ICADE)* in Madrid (2006/08), the American *Harvard University* (2008) as well as the German *ESB Business School* in Reutlingen (2008/10). Finally the academic knowledge is completed with industry expertise gained as a trader trainee at Noble Brasil in Sao Paulo, Noble Argentina in Buenos Aires and Rosario and Noble Resources in Geneva. Noble Group is a Supply Chain Manager and reaches a global scale with its operations encompassing a total of 150 offices in 38 countries. As a Fortune 500 company based in Hong Kong (China) and listed in Singapore, the firm operates regional headquarters in Geneva (Switzerland), London (UK) Stamford (USA), Sao Paulo (Brazil) and Buenos Aires (Argentina). Currently the company employs around 12,000 employees and generates an annual turnover of US$60 billion with a net profit of approximately US$500 million whereas the agriculture products segments represent approximately one third of their business activity besides other sectors such as energy and logistics.[38]

1.3. Structure of the book

After an introduction providing a big picture of biofuels markets within the climate change context, the European Regulation in form of the RED will be presented (Chapter 2). Definitions and background information on the regulatory framework as part of the European Energy Strategy will be given so that the reader will easily be able to follow the methodology of these political objectives. Henceforward, background information on biofuel industry regarding the value chain and market participants will be presented (Chapter 3). Furthermore, commodity trade flows including biofuels and feedstock are illustrated in order to better understand the global interconnectedness of the market. In reference to the RED, sustainability characteristics of biofuels including greenhouse gas emissions and land use change as well as complementary issues like genetically modified organisms and biodiversity are critically discussed. Additionally to the RED, general political interventions in biofuel markets are outlined in Chapter 4, which will be complemented with a case study of Sustainability Certification schemes. Challenges for market participants and criticism on policy frameworks as well as an outlook into advanced generation biofuels and food markets complete the fourth chapter. Finally, the conclusion will critically oversee the European regulation and convey general recommendations to relevant stakeholders in order to pave the way for a sustainable biofuel story in the future.

[38] See Appendix for Noble Company presentation

2. The Renewable Energy Directive (RED)

In the following chapter, a historical background on European policy-making for biofuel will be presented. A general overview of the introduced regulations as well as a more detailed explanation of corresponding contexts will be provided so that a sound basis on the European framework can be gained. This knowledge will serve for a better understanding of the sustainability contradictions and consequences on the biofuel industry that will be outlined and discussed in the following chapters.

2.1. Background on the RED 2003/30/EC on commodity and biofuel trade

The first serious step regarding a common biofuel policy in the EU27 legislation was the Directive 2003/30/EC, which entered into force on 17 May 2003 and the Member States (MS) had to translate into their national legislation by 30 December 2004. The Directive aims to promote the use of biofuels or other renewable fuels for transport. The rationales to bring the Directive into force were the fact that the European transport sector accounts for more than 30 per cent of final energy consumption and keeps expanding as presented in Figure 2.[39] Moreover, an increased use of biofuels for transport is a significant tool among others, which reduces the dependence on imported energy and hence secures a certain energy supply in the medium and long term within the EU27.[40]

Figure 2: Biofuels productions and their contribution to transport fuel consumption[41]

mmt	2002	2005	2006	*CAGR 2002-06*	2010	*CAGR 2006-10*	2020	*CAGR 2006-10*
Transportation	1.1	3.1	5.4	*50.5%*	16.0	*31.3%*	34.0	*20.7%*
Gasoline / oil consumption	286.2	297.2	300.4	*1.2%*	317.3	*1.4%*	349.5	*2.4%*
Biofuels share		*1.1%*	*1.8%*		*5.0%*		*9.7%*	

Additionally, the continuous promotion of biofuels in transport constitutes a step towards a wider application of biomasses which is in line with innovation and research intentions of the EU27 and therefore enables a further development in

[39] European Parliament and Council, 2003.
[40] European Parliament and Council, 2003.
[41] EREC, 2008.

next generation biofuels[42]. Moreover, by the end of 2005, the European Parliament and the Council of the European Union adopted the proportion of biofuels and other renewable fuels, which had been fixed to two per cent based on the energy content of all petrol and diesel for transport purposes offered in the MS markets. This indicatory blending margin was supposed to be raised to 5.75 per cent by all MS in 2010 and nowadays varies between 2.5 and seven per cent as in Figure 3.[43] In compliance with the aforementioned blending targets, MS are eventually responsible for the execution within their national legislative programs. Though, a yearly report to the Commission[44] about nationwide implementations regarding the Directive 2003/30/EC has to be elaborated by all MS.

Figure 3: Development of biofuels mandates in Europe[45]

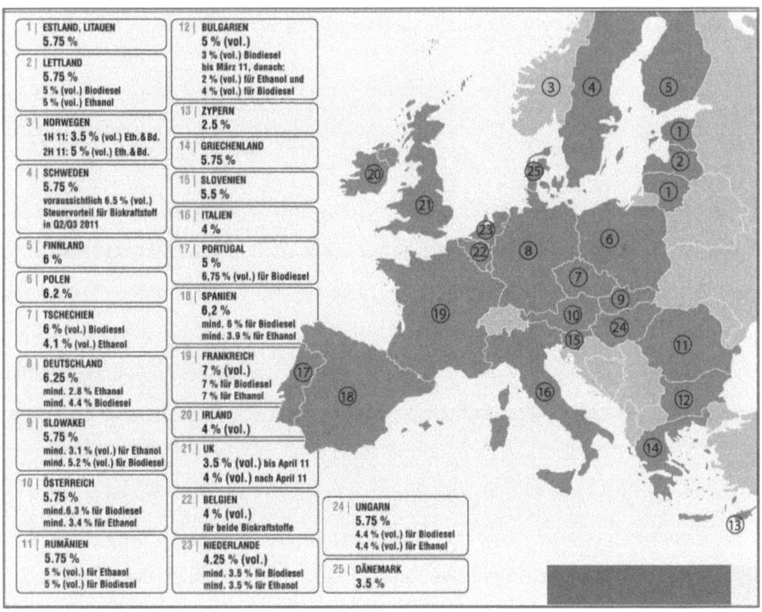

[42] European Parliament and Council, 2003.
[43] European Parliament and Council, 2003.
[44] European Parliament and Council, 2009.
[45] UFOP, 2011.

2.2. The Renewable Energy Directives 2009/28/EC and 2009/30/EC

Based on the RED 2003/30/EC, a new legislation aiming at the GHG reduction and the promotion of renewable energy was adopted by the European Union in 2009. Twenty days after it had been published in the Official Journal of the European Union, the Climate Action and Renewable Energy Policy Package including the Directives 2009/28/EC and 2009/30/EC, under the names Renewable Energy Directive (RED) and Revised Fuel Quality Directive (RFQ) entered into force. This new legislation[46] aimed at the decrease of the EU27 energy dependence as well as a set up of sustainable production practices in agriculture and energy generation with the overall intention to avoid adverse effects of global climate change. In accordance with these directives, MS must commit to the declaration of how they will achieve their targets by December 2010.

This new legislation has become the European 20:20:20 Renewable Energy (RE) Strategy including the following overall targets at the consolidated EU27 level, which are to be achieved by 2020 as summarized in Figure 4:[47] i) a twenty per cent reduction in energy consumption compared to projected consumption for 2020, ii) a twenty per cent share of renewable energy by 2020; iii) twenty per cent reduction of the GHG emission in the EU27 compared to 1990 levels. Within the second target, a ten per cent share of energy from renewable sources in land transportation so that, the new legislation raises challenges and opportunities particularly for stakeholders in the European biofuel sector for the next ten-years. In the biofuel industry there is an urgent need to work out how to cope with new requirements that have been established within the RED and RFQ. Quick reaction by the biofuel producers' side is essential to gain bigger market share offered by the ten per cent target. However the future perspective in biofuel production really depends on how MS implement the new EU27 policy on a national level. So, most MS set out national targets and a roadmap of renewable energy shares to be achieved within energy consumption in transport, electricity as well as heating and cooling by 2020.

Figure 4: Legislation scheme based on EU Directives 2009/28/EC and 2009/30/EC[48]

[46] European Parliament and Council, 2003 and European Parliament and Council, 2009.
[47] European Parliament and Council, 2009.
[48] Own demonstration based on European Parliament and Council, 2003 and European Parliament and Council, 2009

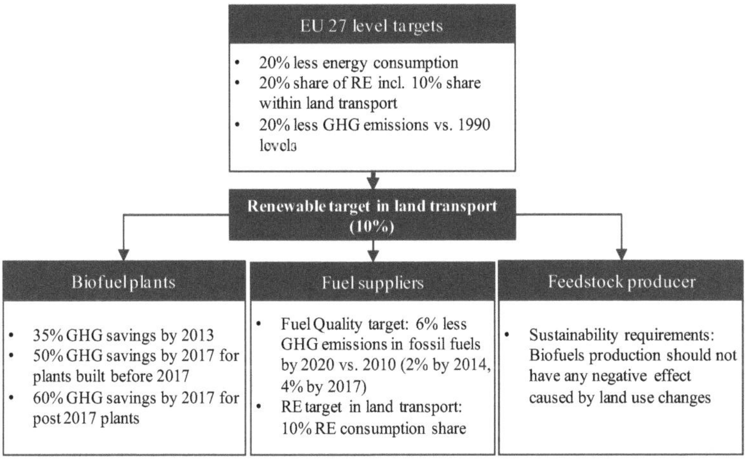

The new legislation requires changes in the behaviour of market actors along the biofuel supply chain to meet the concerning targets. In addition to overall targets set at the consolidated EU27 level, individual objectives have been defined at MS level, where the latter take into consideration the current situation and the growth potential within MS. These Directives set measures, which have significant effects on fuel and energy, supply in land transport, consequently on biofuel production. As aforementioned, one of the targets set by EU indicates a 10 per cent share of renewable energy in land transport, which will inevitably result in increasing demand for renewable energy sources, such as first generation biofuels, next generation biofuels and electric vehicles charged with "green electricity". [49] This target makes the transportation fuel selection more fragmented and result in a competition between different alternatives to fossil fuels. In particular, the second and third targets of the Directive mostly affect stakeholders of the biofuel industry. On the one hand, the ten per cent target of renewable energy in land transport bears an enormous potential for the biofuel sector. On the other hand there is also a certain pressure on the global biofuel market and its actors given that alternative ways to achieve the desired target scenario can be met including electric cars as a renewable energy source. In general, various stakeholders on diverse levels imply the passed legislation in different ways. The first target aims at the achievement of

[49] National Grid, 2011

a 20 per cent improvement in overall energy efficiency by 2020 at the consolidated EU 27 level.

Additionally the second objective comprises a 20 per cent share of RE in total energy consumption, which has to be reached at EU level by 2020 as well. In Figure 5 the current as well as the aspiring RE shares are outlined so that the consolidated EU27 percentage will reach 20 per cent within the next decade. Nevertheless, various MS contribute with different weights as well as higher or lower targets than the 20 per cent of renewable energy in order to achieve this overall goal.[50]. The RED set the national objectives based on historical and current situations as well as potential growth. As in Figure 5, there have been extreme differences for example between Malta (ten per cent) and Sweden (49 per cent). In addition to Sweden, Latvia (40 per cent), Finland (38 per cent), Austria (34 per cent), Portugal (31 per cent) and Denmark (30 per cent) will have to provide a significant contribution to the overall EU 27 goal.

Figure 5: Share of renewable sources in final energy consumption[51]

	share of renewable	target till 2020		share of renewable	target till 2020
Austria	*23.3%*	*34.0%*	**Latvia**	*34.9%*	*42.0%*
Belgium	*2.2%*	*13.0%*	**Lithuania**	*15.0%*	*23.0%*
Bulgaria	*9.4%*	*16.0%*	**Luxemburg**	*0.9%*	*11.0%*
Cyprus	*2.9%*	*13.0%*	**Malta**	*0.0%*	*10.0%*
Czech Republic	*4.6%*	*13.0%*	**Netherlands**	*2.4%*	*14.0%*
Denmark	*17.0%*	*30.0%*	**Poland**	*7.2%*	*15.0%*
Estonia	*18.0%*	*25.0%*	**Portugal**	*20.5%*	*31.0%*
Finland	*28.5%*	*38.0%*	**Romania**	*17.8%*	*24.0%*
France	*10.3%*	*23.0%*	**Slovak Republic**	*6.7%*	*14.0%*
Germany	*5.8%*	*18.0%*	**Slovenia**	*16.0%*	*25.0%*
Greece	*6.9%*	*18.0%*	**Spain**	*8.7%*	*20.0%*
Hungary	*30.0%*	*13.0%*	**Sweden**	*39.8%*	*49.0%*
Ireland	*3.1%*	*16.0%*	**United Kingdom**	*1.3%*	*15.0%*
Italy	*5.2%*	*17.0%*	**EU27**	*11.5%*	*20.0%*

This target can be met either through the use of biofuels and alternative fuels or reductions at the according non-renewable energy production sites. In addition, the FQD set two intermediary targets including a two per cent reduction by end of 2014 and a four per cent one by 2017[52]. Out of these total four per cent, two per cent reduction can additionally be reached with the introduction of environmentally friendly carbon capture and storage (CSS)[53] technologies as well as electric vehicles.

[50] UFOP, 2011.
[51] EREC, 2008
[52] European Parliament and Council, 2009
[53] WRI, 2012

While the first technology has not reached a mature enough level to be introduced into markets,[54] the electric vehicles will play a major role within the upcoming years.[55] That is why at the 82nd *Salon international de l'auto et accessoires* in Geneva, Switzerland, in 2012, most car manufacturers presented their electric, flex-fuel[56] and hybrid vehicle models. This exposition provides usually the status quo of the industry so that upcoming trends can be identified. The other two per cent reduction can be realized with Clean Development Mechanisms (CDM).[57] In order to obtain the six per cent mandatory target only those biofuels can be taken into account, which are produced in a sustainable way as per RED 2009/30/EC[58]. Thus, fuel distributors aim to purchase biofuels that comply with sustainability criteria; and therefore, biofuels refiners require sustainable feedstock. This interconnectedness and complexity of the fuel and biofuel supply chain will be explained in more detail in Chapter 3.2. In Figure 6, a consolidated forecast for total biofuel production within the EU27 shows that the increase from 15 mmt to 35 mmt will cause solid basis for the achievement of the three RED targets. While bioethanol and electricity remain at stable levels, the biodiesel sector will expand at a continuous pace so that this sector will increase in importance for the EU27.

Figure 6: Consolidated forecast for biofuel production in the EU (mmt)[59]

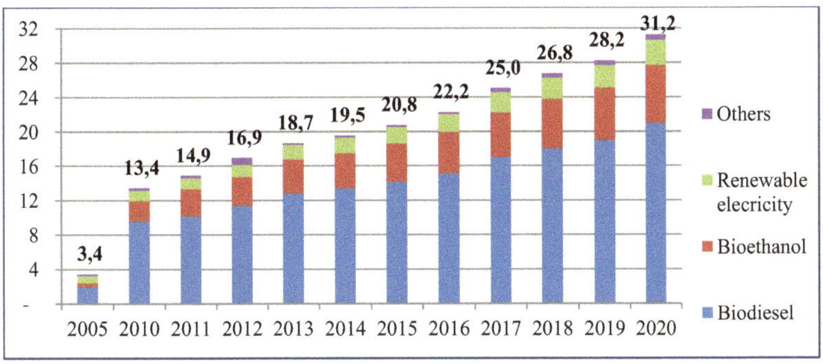

For these sustainability criteria, the EU Commission also defined a sustainability framework setting criteria for biofuel feedstock. Therefore, biofuels in the EU27 cannot be made from raw material obtained from land with levels of high biodiversity. These areas are primary forests, designated for nature protection purposes

[54] REN21, 2011.
[55] Lausanne, 2012.
[56] EERE, 2012.
[57] Quelle zu CDMs
[58] EC, 2009.
[59] UFOP, 2011.

or protection for rare, threatened or endangered ecosystems or species and grasslands that represent a high value of biodiversity. Furthermore, neither raw materials obtained from land with high carbon stock, that is wetlands, nor continuously forested areas, “land spanning more than one hectare with trees higher than five meters and canopy cover between 10 per cent or 30 per cent, or trees able to reach those thresholds in situ”,[60] nor peat land can be considered for biofuel production. According to the RED, voluntary schemes have been established to which stakeholders along the biofuel supply chain can commit and get a certification. These voluntary schemes have to be accepted by the European Commission in order to enable the market participant to label their material as sustainable. At present, there are seven approved schemes with different levels of commitment that can be chosen by market players, which will be presented in more detail in Chapter 4.3. These sustainability labels or certificates focus on biodiesel and bioethanol exclusively and enable these products to be traded at the European market. In the case of biodiesel, Figure 7 demonstrates what type of feedstock is used for the production in Europe. Especially the increasing shares of PO and SBO represent a thread for the local agriculture industry that depends on the rapeseed oil production. Due to the oil imports from South America and South East Asia triggered by very competitive prices, the European production of vegetable oil has been recently stagnating. Regarding the vegetable oils, only SBO is covered with sustainable certificates whereas for the other oils, the European commission has not approved one yet.

Figure 7: Types of biodiesel feedstock used in the EU27[61]

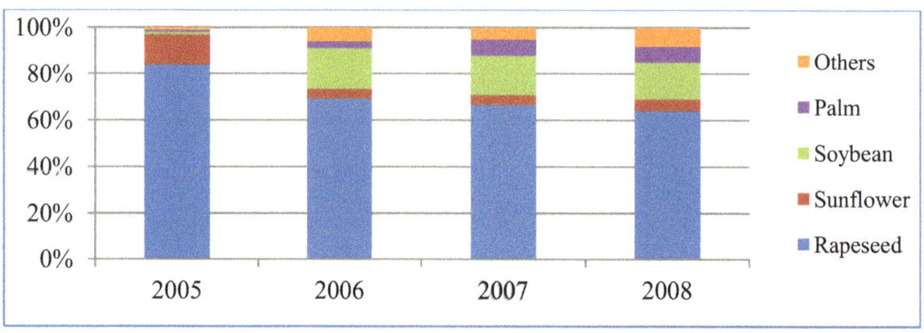

[60] FAO, 2007.
[61] AEBIOM, 2011.

Nonetheless, this stagnation of biodiesel production is not driven by a lower demand. As Figure 8 indicates, the EU27 consumption has been rising over the last decade and is predicted to increase up to approximately 22 mmt by 2020. That is why, demand and supply have to be analysed and evaluated separately. In Chapter 3.5, the commodity trade flows including import and export of major oilseeds, vegetable oil and biodiesel products will be explained in more detail.

Figure 8: Biodiesel consumption in the EU within the land transport sector (kmt)[62]

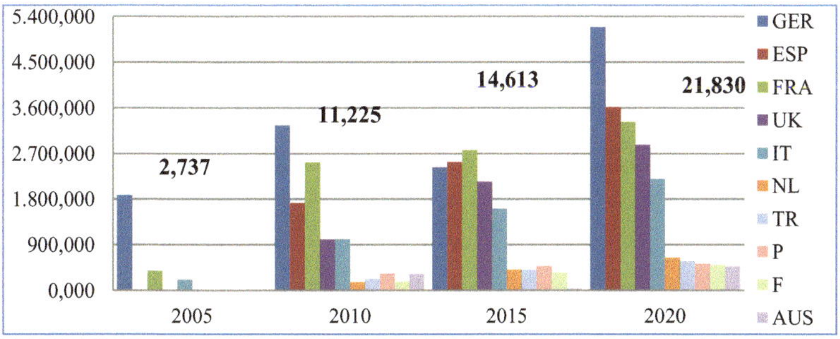

By 2020, the ten per cent target of the RED will serve as key instrument within the EU climate change mitigation strategy. Transport is indeed the only EU economic sector where GHG emissions are growing unabated. Additionally, the target will reduce the EU's dependency on fossil fuel sources and oil imports, therefore contributing to strengthening the EU energy security. In this perspective, it is significant to guarantee the full efficiency and scope of the ten per cent is maintained. This means that the objective should incentivize the consumption of those technologies that will make a significant contribution to an increased use of renewable energy in the transport sector by 2020. Any diverging approach would not be compatible with the overall RED objective.[63] As Figure 8, MS have clearly indicated in their National Action Plans that biodiesel will contribute the largest part of the ten per cent target for 2020 at the EU27 level (66 per cent of the target should come from biodiesel, 23 per cent from bioethanol). Electricity and biogas are expected to represent only a marginal contribution.

[62] UFOP, 2011.
[63] EBB, 2011.

In order to verify biofuels' compliance with sustainability criteria a mass balance system method (MBS)[64] is applied. The use of MBS allows the European Commission to create more transparency as market players must track the quantity of sustainable materials and products proving the fulfilment of the sustainability criteria. However, thanks to the MBS, feedstock and different sorts of biofuel can be easily identified, for instance a B100 biodiesel that is made from 40 per cent RSO, 40 per cent animal fats and 20 per cent from waste/used cooking oil. Therefore only biofuel or a proportion of feedstock actually meeting the sustainability threshold value[65] can be taken into account of the MBS. Crucial for the biofuels and the emission targets to be achieved by MS are the overall GHG emissions. Each biofuel possess a specific GHG emission values collecting all the GHG emissions occur during their life cycle. The European Commission provides data about GHG emissions of different biofuels depending on their individual feedstock and production processes in the RED Annex.[66] Chapter 3.6 will particularly explain how GHG emission savings are calculated based on a lifecycle assessment (LCA).[67]

[64] The mass balance system method will be explained in detail in chapter 4.3.
[65] The minimum GHG reductions are 35 per cent until 2017, 50 per cent between 2017 and 2018 and 60 per cent after the year 2018.
[66] USDA, 2009.
[67] Teulon, 2011.

3. The biofuel industry

In this chapter, a general overview of the industry and the role of different participants along the biodiesel supply change will be presented, including crop producers, crushing plants, biodiesel refineries, and fuel suppliers before the focus will be laid on more detailed insight into different feedstock and biodiesel production. Thus a profound basis will be granted leading to a sound understanding of the sustainability criteria related to climate change and policy making that will be covered in Chapter 4. In order to finally grasp the interconnectedness of the biofuel industry and worldwide networks of supply chain management, global commodity trade flows will be outlined as well as the effects caused by the RED.

3.1. An overview of the biofuel industry and its players

There are a number of different actors involved in the value chain promoting biofuels. According to McKinsey, these actors are asset owners, general market participants and product and service providers. Whereas these asset owners are usually agribusinesses, petrol industries, gasoline blenders, chemical industries, plant operators and farmers, market actors invest in the biofuel industries. The product and service providers involve seed and fertilizer suppliers as well as engineering and agriculture equipment companies.[68] The biofuel industry is accelerated by a large wave of investments from different business sectors. Some of the main investments come from multinational agricultural commodity companies as Archer Daniels Midland (ADM), Glencore, Cargill and Noble Group, who are making rather substantial investments into agrofuels as well as companies that have an interest in sugar trade, palm oil and forestry.[69] From the energy sector large corporations as Beyond Petrol and Mitusi are investing in biofuels. There are also the investments from oil companies that have direct relations to the national governmental agendas as Petrobrás in Brazil and Petrochina or Philippine National Oil Company.[70] The most substantial investments in biofuels are believed to come from the financial sector, some of the main sources of globalized

[68] McKinsey Global Institute, 2011.
[69] ADM, 2012 and Noble group, 2012a.
[70] petroliumfirmen, etc sourcen

capital as banks, such as Barclays, Morgan Stanley and Goldman Sachs are investing in biofuel industries. Investments into biofuel industries are also coming from some of the wealthiest individuals in the world, billionaires such as George Soros, Vinod Khosla and Bill Gates. Richard Branson is the owner of Virgin Group and therefore directly connected to Virgin Fuels. They all have large investments into biofuels.[71] Governments and international monetary institutions including the World Bank and regional development banks also provide support to this industry in terms of subsides, tax releases, carbon-trading schemes and soft loans-to sustain the economic viability of biofuels. In general, these investment flows are a restructuring effect on agribusiness, which re organize and strength transnational structures, connecting the land owning elites of the South "with the most powerful corporations of the North". It seems that biofuel industries are "being managed by transnational corporations and absorbed into their profit strategies and expansion plans".[72] The politically set targets for blending requirements are guaranteeing a demand for business, that in addition to governmental subsides make it an interesting business opportunity. The most crucial factor is the price for feedstock as the major production cost driver arising to 60 per cent.[73] The feedstock price "can make or break a biofuel production".[74] Prices for the biofuel industry is competing with other markets, in particular the food market, as one is using the same crops and lands. In Chapter 4.6, the focus will be expanded on food price volatility and hunger. When the use of biofuels will increase, feedstock prices will also increase and have an impact on supplies. High feedstock prices are risky for any business. Therefore there are many options to lower production costs. In order to ensure low costs many commodity companies are controlling both the production and supply of the feedstock through integrated supply chains of biofuel networks that integrate the whole production process from seeds to shipping and distribution. Another strategy for keeping down production costs is relocating facilities to cheaper countries or via enhanced biotechnology. Genetic Modification or improvement of Organisms (GMO) is attractive in terms of yield increases and reduction of agricultural inputs.[75]

[71] Daab, 2012.
[72] GRAIN, 2007b.
[73] McKinsey Global Institute, 2011.
[74] GRAIN, 2007a.
[75] Editora Gazeta, 2011.

All over the globe, governments are re-examining their energy alternatives, in particular in search for lower carbon emissions in terms of cleaner fuel alternatives and technologies. Peaking oil prices as they have developed during the last decade, the expected depletion of fossil fuels, concerns for global warming and subsequent climate change are some of the main reasons that explain the growing global biofuel industry. A concrete picture and clear indication of the interest in cleaner energy is provided by the amount of investments generated. During the past five years, worldwide investments in renewable energies (RE) have been particularly occurred within the biofuel sector as shown in Figure 9. Over US$70 bn have contributed to an enormous increase in global biofuel expansion. Furthermore, the United States has been investing over US$ 100 bn in clean energy by end of 2010, compared to US$38 bn in 2005 or even to US$5 bn ten years ago.[76] Another major influential power, China, seems to follow in the same direction, as it has announced that it will invest US$ 187 bn in clean energy by 2020.[77]

Figure 9: Development of global financial new investments in RE excl. wind (bn USD)[78]

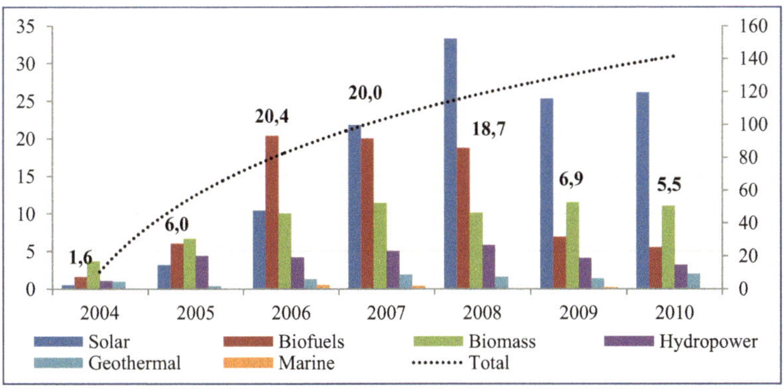

Biofuels have become an integral part of this shifting investment environment, driven by the growing number of governments worldwide that implement regulations and frameworks such as blending targets and by the estimated future investments in renewable energies.[79] Moreover, triggered by heavy private sector

[76] IAB, 2007
[77] IAB, 2007
[78] Own demonstration based on REN21, 2011, UFOP, 2011 and Ragwitz, 2005.
[79] IADB, 2007.

investments and favourable regulations, Germany has for instance developed a particularly strong biofuel sector that generates a yearly industry turnover of over US$ 4.5 bn (Figure 10). Therefore, this industry partially driven by capital intensive assets and high inventories in Germany is approximately 40 per cent larger than the comparatively established wind and photovoltaic (PV) ones. During the last decade, biofuels generally have become viewed as a serious alternative to conventional fossil based fuel for transport thanks to the simple adaptability of motors and existing distribution infrastructure. Furthermore, the most efficient producers have even prices competing with the oil industry as in the case of Brazil and Argentina.[80]

Figure 10: German industry turnovers related to renewable energy facilities in 2011 (in million EUR)[81]

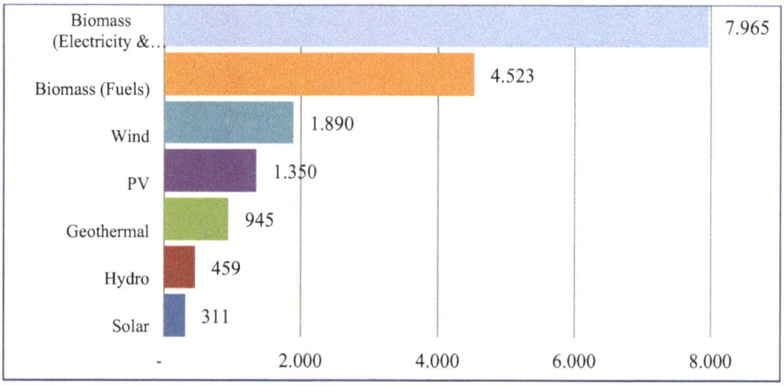

Generally, to a certain blend, both bioethanol and biodiesel can be used in conventional engines. Since an engine modification is required for pure biodiesel (B100), diesel mixtures work up in form of biodiesel blends as for instance five (B5) and ten per cent (B10) biodiesel are added to conventional crude oil based diesel.[82] That is why most of the biodiesel distribution is through blends, which are encouraged by government regulations as aforementioned in the RED overview.[83] Moreover, the Brazilian president plans to increase the margins up to ten per cent by 2015 as well and therefore, it would be in line with European regulation. The general trend over the last decades represented an increase in both bioethanol than biodiesel production. Bioethanol is mostly produced and con-

[80] Dufey, 2006 and USDA, 2011a.
[81] BMU, 2012.
[82] ESMAP, 2005.
[83] ABIOVE, 2012a.

sumed in North and South America, whereas the European market is the world-leading producer of biodiesel and consumes most of it domestically. Focusing on the European Market, Figure 11 provides a big picture of biofuel consumption; Germany, France and Spain represent currently the major producers. This hegemonic position will probably move towards South America, in particular driven by Argentina and Brazil where biodiesel production has still a huge potential to expand and which have been catching up recently due to strategic capital investments by the private sector.[84]

Figure 11: Biofuels consumption in EU27 in 2010 (kmt)[85]

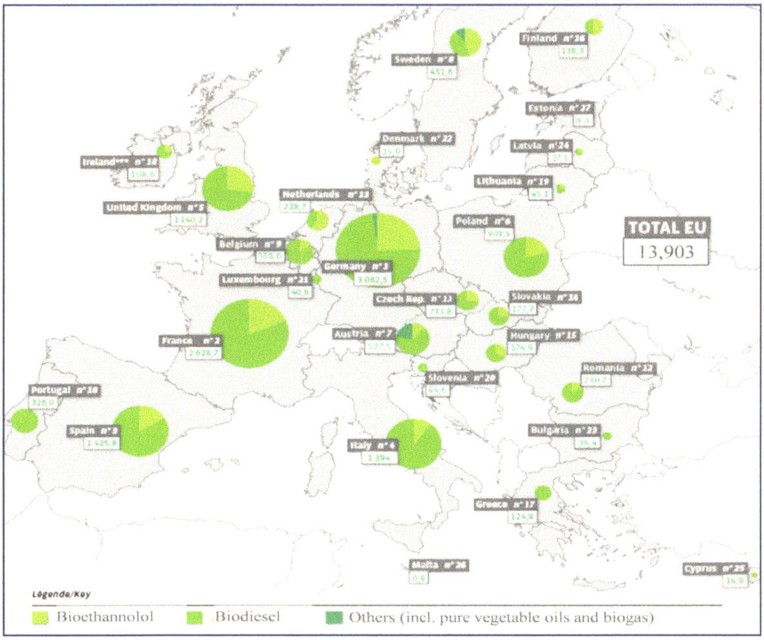

As Figure 1 demonstrates, more biodiesel than ever before in history is sourced from abroad so that global trade spans a global network, so that global biodiesel production grew exponentially and surpassed 16 mmt in 2010 compared to less than one million metric ton ten years ago. As aforementioned in Figure 1, biodiesel production is mainly concentrated in Europe with over 9mmt as well as in

[84] Dufey, 2006.
[85] EurObserv'Er, 2011a.

Latin America, where Argentina produces around 2.3mmt and Brazil 2.2mmt as well as Colombia and Peru with 0.4mmt and 0.1mmt respectively.[86]

Figure 12: Development of world biodiesel production between 2000 – 2010 mmt)[87]

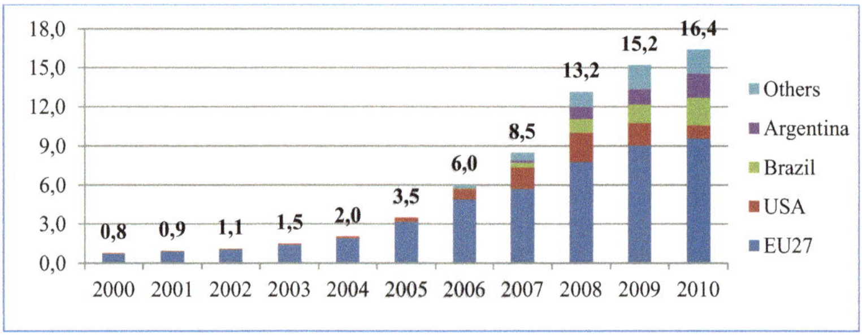

3.2. The biodiesel value chain and different types of feedstock

Generally biofuels can be broken down into three different types; in liquid form there is bioethanol and biodiesel and in gas form there is biogas. All three fuels are made of biological and agricultural resources characterizing them as first generation biofuels. Bioethanol is mainly produced with sugar plants and cereal crops, mainly corn and sugar cane, but also beet, cassava, wheat and sorghum. Raw materials used for biodiesel production are oilseeds, for example rapeseed, sunflower, soy, palm, coconut or jatropha. In detail, bioethanol in North America is mainly produced with corn while in South America the major feedstock stems from sugar cane. Biodiesel in South America is mainly produced with SBO whereas in Europe, and in particularly, in Germany and France, the refineries operate with RSO as principal feedstock (65-75 per cent market share of RSO).[88]

There is quite a great difference in oil yields of the different plants. The most oil producing plant is palm oil (PO), which yields 5,550 l/ha,[89] in comparison to soybean oil (SBO) that reaches only 420 l/ha. The mainly in south east Asia, and especially Indonesia and Malaysia, produced PO is the largest existing source of vegetable oil globally and yields about 13 times more than soybeans. The variations between oil yields depend on the content of oil that a plant gener-

[86] BioDieselBr.com, 2011.
[87] Own demonstration based on EBB 2010, 2011 and UFOP, 2010, 2011a.
[88] Noble Group Research, 2012b.
[89] Mukhtar and Singhal, 2012.

ates and not by how much volumes can be produced per hectare. On average one litre of vegetable oil produces approximately one litre of biodiesel.[90] For instance, a comparison between soybeans and sunflower seeds in Argentina leads to the result that soybeans generate higher yields with approximately 2,200kg/ha compared with 1,700 kg/ha for sunflower seeds. However, sunflower generates almost double the amount of yield in litre per hectare. This is due to the difference of oil content in the plants. The oil content for soy is estimated to be 19 per cent of the plant, whereas the sunflower seeds' oil content is between 30-40 per cent. The choice of feedstock is primarily determined by the climate and cropping conditions of the region as well as being a function of the prevailing agricultural production of a country and its vegetable oil industry, as an established production chain, infrastructure and markets for vegetable oil production already exists.

As aforementioned, several feedstock serve for the biodiesel production, including more than 350 oil-bearing crops identified that contain fatty acids, such as soybean, sunflower, rapeseed, cottonseed, castor beans and peanuts. Nonetheless, animal fats and tallow, especially recycled greases and used cooking oils have gained increasing attention in recent years, in particular because of the regulation and GHG emission calculations.[91] Most common feedstock for biodiesel production are rapeseed in the EU, soybeans in South America [92] and palm oil in South East Asia.[93] In the EU over 60 per cent[94] of biodiesel is produced from rapeseed oil, in the US, Brazil and Argentina SBO dominates, and regarding Southeast Asia palm oil is the main source of the production.[95] Rapeseed produces about 1,100 litters of biodiesel per hectare, and in case of high-yield rapeseed the yield can reach 1,400 litters per hectare while soybeans have lower oil content than rapeseed. Therefore soybeans produce about 400 litters biodiesel per hectare. Moreover, sunflower generates approximately 700 litters whereas the potential biodiesel yield of palm oil is 2,409 litters per hectare.[96]

[90] Depending on the quality, temperature and biodiesel feedstock, SBO's average density is 0.93 t/m^3, RSO: 0.92, Gasoline 0.75, Diesel: 0.83 and B100 0.82-0.9,that is, RME 0.88 based on Kaltschmitt, M, 2009.
[91] European Parliament and Council, 2009.
[92] Van Gerpen, 2004.
[93] Vasquez, 2012.
[94] Oil World, 2009.
[95] De Ridder, 2005.
[96] Monsanto, 2012 and Vasquez, 2012.

Going back in history, soybeans are originally from Southeast Asia and had been imported to America where most production takes place. Nowadays up to 90 per cent are mainly produced in the US, Brazil and Argentina, which are the three world leading soybean producers whereas the US is the largest producer, Brazil the largest soybean exporter and Argentina the largest Soybean complex, i.e. soybean meal and soybean oil exporter. Soybeans are grown for its oil and meal that contains a high level of proteins. Thus, soybeans have become a major commodity over the last decades supplying numerous different industries as demonstrated in Figure 13. After harvesting the soybean, the beans are crushed so that three different products can be gained. First the soybean meal (SBM) which contains a high protein level, second soybean pellets, which basically contain the hulls of the soybean and compose of a high level of fibres and third the soybean oil that is extract from the soybean. While the oil can be used for various industry destinations, the meal and the pellets are mainly used for animal feed.[97] Nowadays, 85 - 90 per cent of the total soybean production is crushed while ten per cent are used by the food and feed industry as whole beans. An inconsiderably small part of the total crop is consumed by people as whole soybeans, in particular non-genetically modified organism (GMO), for instance in Japanese or organic cuisine. Depending on the origin of the soybeans and on the technology of the crushing plant vary the yield of oil content gained by the crushing process. On average, crushed soybeans provide approximately 79 per cent soybean meal and 18 per cent of soybean oil whereas the latter one is the much more valuable product.[98]

[97] Editora Gazeta, 2011.
[98] Industry consensus, Oil World 2009 and WWF, 2003.

Figure 13: Soybean and soybean products production flow[99]

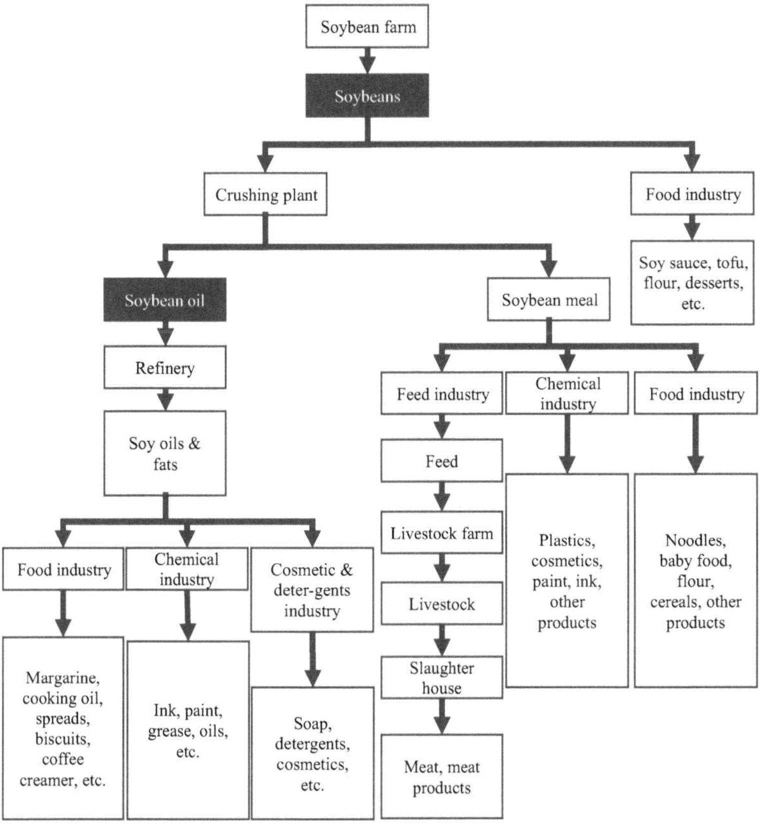

Figure 13 provides an overview of the corresponding use in industries worldwide. Especially, in the food industry soybeans are used to produce soy sauce, tofu and other kinds of meat substitutes. SBO is mainly used as table oil, but also as an ingredient in mayonnaise, margarine, pastries and snacks. In general, the main driver for crush is the SBM demanded by the feed industry to produce primarily compound feed for the livestock industry.[100] As it is rich in proteins and low in raw cellulose, it is especially suitable for single stomached animals such as hogs, broiler and hatches.[101] Additionally, SBO is also connected to the cosmetic industry as ingredient for beauty care products, detergents and soap. Finally it is

[99] Own demonstration based on WWF, 2003 and Noble Argentina, 2011.
[100] BELV, 2012.
[101] Noble Group Research, 2012a.

the chemical industry that purchases vegetable oil for multiple purposes including paint and lacquer production as well as for soy diesel and soy ink.[102] As Figure 14 illustrates a concrete example in Germany, RSO is the major feedstock for biodiesel production and generally the refinery facilities demand approximately 55 per cent of total vegetable oil consumption while the food and oleochemistry industry represent comparatively minor shares.

Figure 14: Biodiesel blend feedstock, oil consumption and biodiesel use in Germany in 2011[103]

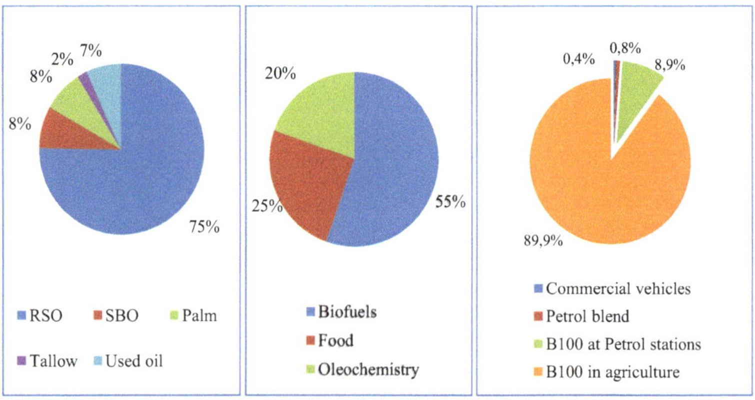

The SBO continues as major feedstock along the biodiesel supply chain towards the biodiesel refinery. There are different steps along the biodiesel value chain starting from the farmer that produces the crop as feedstock and sells it to the exporter either via direct sales to the exporters or through elevators, which are mainly larger corporations collecting the physical material and storing it in order to resell it. These elevators either purchase the merchandise or only provide a storage service so that the ownership of the soybean does not necessarily changes. With or without elevators, the beans are mainly sold through broker houses that consolidate offer and demand within the national and international market, that is, the physical goods for instance will be sold either to local crushers or export companies being the main actors along the biodiesel value chain (Figure 15).[104] There are various companies searching for an integrated approach within the Supply Chain including big agriculture commodity trading companies like

[102] WWF, 2003.
[103] Own demonstrations based on Noble Group Research 2012b, UFOP, 2011b and FNR 2011a.
[104] Simantob, 2011.

Cargill, *ADM*,[105] *Glencore* or *Bunge*. Heavy investments or combined oil extraction plants with biodiesel refineries (transesterification facilities) so that companies can become vertically integrated, achieving higher profit margins. Below average market conditions, the transesterification, i.e. the biodiesel production provides higher margins in comparison with the vegetable oil export so that a crushing plant's capacity can reach higher utilization.[106] Moreover, vertical integration along the value chain provides necessary market presence leading to increased information about current trends and trade flows as well as creates flexibility on the supply side. The company itself can decide upon market prices whether to produce oil or biodiesel and in which proportion depending on current market levels. This flexibility also provides bargain power not only towards farmers and suppliers that are usually much smaller players in the market but also towards other actors along the supply chain including shipping agents, ship owners and surveyor companies.[107]

Figure 15: Biodiesel Value Chain

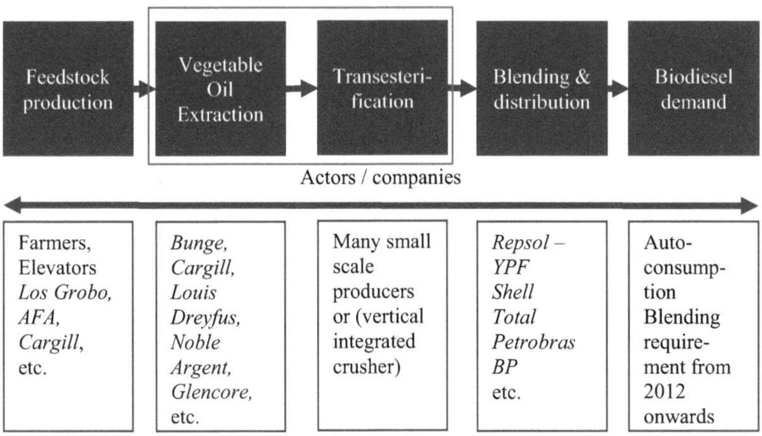

The crushers and refiners are directly linked to farmers and the agricultural sector as most of them grant credits to farmers to cover their cost during production periods. For a soybean farmer, more than 40 per cent of his operating costs are fertilizers so that usually most of these credits are not in monetary terms but in

105 ADM, for instance, operates as part of their global commodity activity in Germany three biodiesel plants with approximately 1 mmt per year refinery capacity as of EurObserv'ER, 2011a and ADM, 2012.

106 Noble Group Research, 2012c and Gallo, 2011.

107 Montenegro, 2011 and Noble Argentina, 2011.

physical form such as fertilizer or seeds.[108] Thus, firms engaged in the vegetable oil extraction make sure that through these state of dependences, they secure their steady supply in the following harvest season.[109] Particularly at low prices, farmers tend to be reluctant to sell the beans and can cause pressure to the market in form of origination shortages. These shortages can be dangerous for crushing plants as they rely on a high utilization rate in order to cover immense fix and operating costs. Over the last decade, the vegetable oil industry has experienced a continuous increase in milling capacity and vegetable oil extraction especially in Argentina as well as the expansion of biodiesel refinery capacity in Europe, and particularly in Germany and Spain as in Figure 16.[110] As raw material in terms of soybeans is available and remains comparatively cheap[111], and there are enhanced milling capacities, vegetable oil producing companies have a strong interest in large scale production of biodiesel and therefore a secured demand. In Latin America and in particular Argentina, their current crushing activities are already export oriented, e.g. strategically located milling facilities at the riverside or even soybean crushing and transesterification which are integrated into port facilities in order to directly load the processed oil into oil tankers that leave mainly for European and southeast Asian markets. This export orientation also allows the vegetable oil producers to preserve a strategic position in the value chain because biodiesel could be exported as a neat fuel for blending and distribution abroad, i.e. serves as a direct supply for petrol companies.

Figure 16: Biodiesel production (PROD) and production capacity (CAP) in selected countries[112]

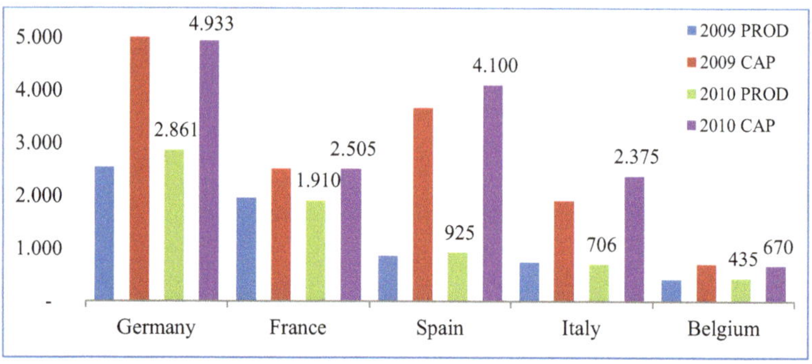

[108] The value chain could be extended to seed or fertilizer producers such as *Monsanto* and *Bayer Crop Science*
[109] De Souza, 2011.
[110] Noble Group Research, 2012c.
[111] Almada, 2006
[112] Own demonstration based on EBB, 2011.

3.3. Feedstock production, crushing plant and oil production

The quantity of oilseeds, which crushing plants are able to process, highly depends on the type of seeds as well as the facility itself. In particular in Argentina where the crushing industry is mostly export oriented, the crushing capacity reaches up to 6,000 to 10,000 tons of soybeans per day. Moreover in Germany there are huge discrepancies between enormous industrial plants on strategic places like the ADM crushing plant in Hamburg that is specialized on feedstock imported from overseas and has a six kmt per day capacity,[113] while smaller family owned crushing plants in the south cannot compete regarding volume with these big ones. These have far smaller capacities, between 30 and 300 tons per day and are flexible in terms of feedstock.[114] In general, most of the crushing plants in Europe are capable to produce multiple feedstocks given that the seed-to-oil conversion ratio ranges between 0.19 tons of oil per ton of beans (soybeans) and 0.44 mt of oil per metric ton (mt) of seeds (SFS).[115] Nevertheless, the proportion of various oilseeds in terms of capacity utilization is hard to track as the industry is quite fragmented and non-transparent. The big players and integrated supply chain manager operating large crushing plants will run the crushing plants non-stop as the fixed costs are very high and the margins are generally low so that the high volume provides the satisfying economic result. In contrast to soybeans, rape and sun seed, palm oil can only be transported as oil and has to be crushed immediately so that the two major producing areas, Indonesia and Malaysia export the liquid oil in huge volumes. There is an elevated potential for palm oil as biodiesel feedstock due to a comparatively high yield per hectare.[116] Furthermore, the rapeseed is the most prevalent feedstock for biodiesel production in Europe, it has the largest stake in the crushing with over 21 mmt of RSS, twelve mmt of SB and around seven mmt of SFS were crushed in 2011.[117]

[113] ADM, 2012.
[114] Hass, 2011 and FO Licht, 2010.
[115] Gallo, 2011.
[116] Noble Group Research, 2012b.
[117] Numbers based on Oil world, 2009 Noble Group Research 2012b and industry consensus.

As part of most of the crushing plants, an oil extraction refinery facility produces edible oil for food and industrial purpose as well as feedstock for biodiesel production.[118] Refined vegetable oils provide the best raw material to produce biodiesel since the conversion ratio of pure triglyceride into fatty acid methyl ester (FAME) is elevated and reaction time is comparatively short. Nevertheless, waste or used cooking oil (UCO) can also be suitable for biodiesel production, particularly when it is properly collected in terms of GHG emissions of collection logistics.[119] However, UCO requires a pre-treatment before it can be used as feedstock which includes filtering and heating processes.[120]

3.4. Biofuel production and chemical background of transesterification

Biofuels are produced in two phases. First there is the cultivation of feedstock by conventional agricultural production as well as the extraction of either vegetable oil via crushing and extraction plants or the fermentation of sugars. Second, after having the raw material, it is converted to biofuel in a refinery, whereas in the case of biodiesel via transesterification and for bioethanol through a distillation processes.[121]

Figure 17: Biodiesel and Bioethanol[122]

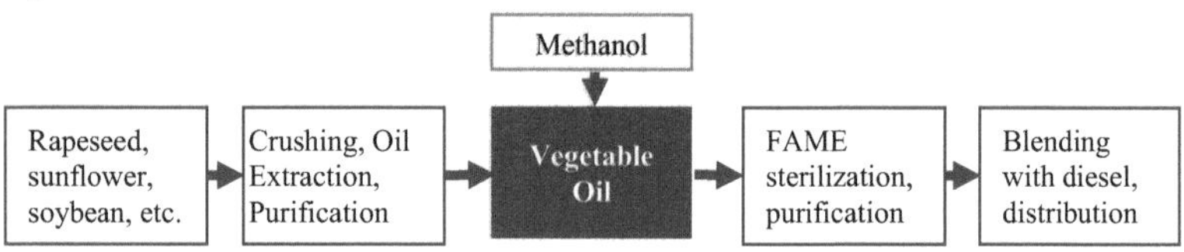

After crushing the oil seeds physically and separating the hulls from the meal that is pressed in into flakes, a chemical solvent is used, as for instant, hexane to extract the oil from the soybean flakes. Usually, the soybeans are cracked, checked for moisture content, pressed into flakes and then treated with hexane as the chemical solvent within the extraction process. The resulting crude soybean oil product is degummed, refined, blended depending on what kind of application

[118] Knothe, 2005.
[119] Demirbas, 2008.
[120] Pazouki, 2010.
[121] Knothe, 2005.
[122] Dufey, 2006.

and in some cases hydrogenated.[123] The SBM, the larger part of the production, is sold as animal feed.[124] After the extraction of the oil, the valuable hexane will be separated again from the oil in order to be reused. Then, methanol is added in order to chemically react with the oil to generate biodiesel and glycerine. Finally, there must be only a separation of the two by-products to isolate the biodiesel. After the crush and oil extraction, the vegetable oil is converted into biodiesel within the biodiesel plants in which the high viscosity of vegetable oils are reduced in order to use the fuel in common diesel engines without operational difficulties. Of the four common practices to decrease viscosity, namely blending with petro diesel, pyrolysis, micro emulsification, and transesterification[125] only the last one is appropriate for biodiesel production.[126] Figure 18 below shows the simplified input-output scheme of biodiesel applying the transesterification process.

Figure 18: Schematic biodiesel production

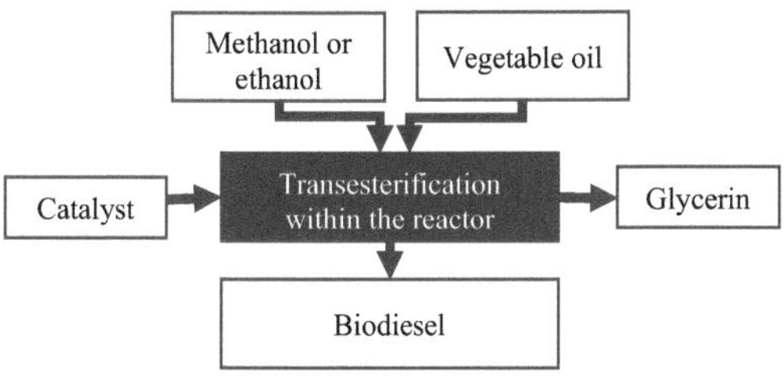

Furthermore, Figure 19 illustrates a schematic overview of the biodiesel production in more detail referring to a basic biodiesel plant consisting of reactors, pumps, settling tanks, centrifuges, distillation columns, and storage tanks. In the reactors, the transesterification takes place in which the reactants are combined and agitated for one hour at approximately 60° Celsius.[127] Plants with lower capacity are constructed with batch reactors whereas larger plants are equipped with continuous flow processes. In a batch reactor reactants are added in one step so that the chemical reaction can occur in a closed space.[128] Moreover, an excess

[123] Noble Argentina, 2011 and Montenegro, 2011.
[124] IADB, 2006.
[125] For the purpose of simplification, the four different transesterification: Base-catalyzed transesterification, Direct acid-catalyzed transesterification, Conversion of the oil into its fatty acids and then into biodiesel (enzyme catalyzed) and Non-catalytic transesterification of oils and fats will be summarized as one process.
[126] Knothe, 2005.
[127] Tomei and Upham, 2009.
[128] Van Gerpen, 2005.

of methanol, with a 20 to one ratio, is necessary given a water accumulation. This amount of additional methanol is required to meet the glycerine standard[129] for fuel-grade biodiesel. Then the unused alcohol has to be recovered and recycled back into the process in order to economize operating costs and limit overall environmental impacts in terms of GHG emission values. In contrast to the batch reactor, the continuous reactor can either be a Continuous Stirred Tank Reactor (CSTR) or a Plug Flow Reactor (PFR). CSTR allow operating with less alcohol within two steps. First, approximately 80 per cent of the methanol and catalysts are added to the oil, and then the stream flows from the reactor to the glycerine removal while most of the water content is removed during the first step.[130] In this second stage, the remaining 20 per cent of the alcohol and catalysts are removed.[131] In a PFR, reactants are introduced from one side so that the composition changes given that the stream moves in plug flow through the reactor.[132] Then, glycerine is separated from methyl esters, which normally occurs either through settling within the tank or by centrifuging. An excess of methanol slows down the separation process. The addition of water improves the separation step so that finally two product outputs, in form of fatty acid methyl esters (FAME) and glycerine are generated. The FAME continue with a neutralization step, during which acids are added to the biodiesel to neutralize any residual catalyst and split up soapy substances that have been formed during the chemical reactions. These acids then form soluble salts and free fatty acids (FFA).[133] After the neutralization, the methyl esters pass through a so-called methanol stripper, in form of either a vacuum flash process or a falling film evaporator, during which salts are removed by water washing. This washing intends to separate remaining catalysts, soaps, salts, methanol or free glycerine from the biodiesel. The FFAs remain and a vacuum flash process removes the water used for the washing so that biodiesel production reaches the final process step. As the glycerine obtained within the separator represents approximately 50 per cent of the total glycerine, further filtering is required. So, the biodiesel is usually sold to a glycerine refinery in order to achieve a purity of 85 per cent and commonly even reaches up to 99.7

[129] 0.24 per cent of total glycerin is permitted in Biodiesel following the ASTM D6571 standard of Free and Total Glycerin Specifications for Biodiesel, as of McCurry, 2007.
[130] Demirbas, 2008.
[131] Van Gerpen, 2005.
[132] Demirbas, 2008.
[133] Free Fatty Acids (FFA) are

per cent. Furthermore, the methanol removed from glycerine and methyl ester output, has a significant value and can only return to the process after passing through the distillation columns in which the remaining water content is extracted. During the aforementioned pre-treatment processes, acid catalysts are added to feedstock with higher FFA levels in order to reduce FFA level down to five per cent content maximum. On average, up to five per cent FFAs, the chemical reactions can still be catalysed with an alkali catalyst; nonetheless, additional catalysts might be required. Especially in case of higher FFAs, soapy substances hinder the separation process and contribute to emulsion formation during the water wash.[134]

Figure 19: Schematic process flow for biodiesel production[135]

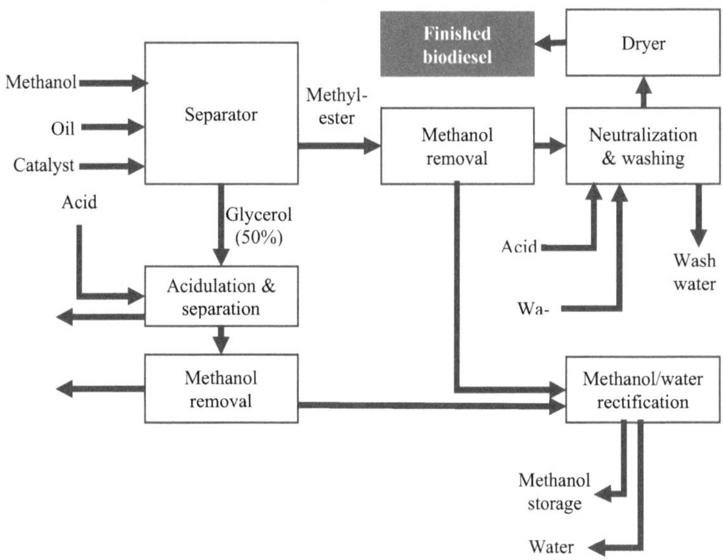

The core reaction within the biodiesel production is the transesterification, which is a chemical reaction of vegetable oil or animal fat with a monohydric alcohol like methanol or ethanol in order to form methyl-esters and glycerine. The transesterification aims to lower the viscosity of the oil or fat. This reaction depend on different circumstances such as reaction temperature and time as well as the molar ratio of glycerides to alcohol, FFA, the water content of vegetable

[134] Zylla, 2012.
[135] Own demonstration based on Knothe, 2005 and Kabath, 2012.

oils or animal fats and finally the type and amount of catalysts.[136] In the "Handbook of Biodiesel", van Gerpen outlines that a reaction can vary in terms of time between one hour at approximately 60 grad Celsius or four hours at a lower temperature of approximately 30 grad Celsius.[137] Furthermore, FFAs and the water content inhibit the reaction so that it is of utmost importance that feedstock complies with industry quality standards. The average yield of methyl esters amounts approximately 67-86 per cent for crude vegetable oils and 93-98 per cent for refined oil, particularly caused by FFAs representing up to 7 per cent within crude oils. Moreover, crude oil comprises other substances such as phospholipids that further lower the quality of the oil.

Generally, commercial biodiesel producer apply a base-catalysed reaction using sodium or potassium hydroxide. In spite of the acid-catalysts utilization that pre-treats feedstock, converting the higher FFA values into esters, the reaction rates for the triglycerides conversion into methyl esters are comparatively slow.[138] Finally, the transesterification process can be summarized in form of the chemical equation below (Figure 20). R1, R2, and R3 are longer hydrocarbon chains or so called fatty acid chains.[139] Depending on the feedstock, different types of oils or fats are represented with distinctive sorts of fatty acids so that the difference among the chains lies in their length, degree of unsaturation or presence of other chemical functions.[140]

Figure 20: Chemical equation of transesterification[141]

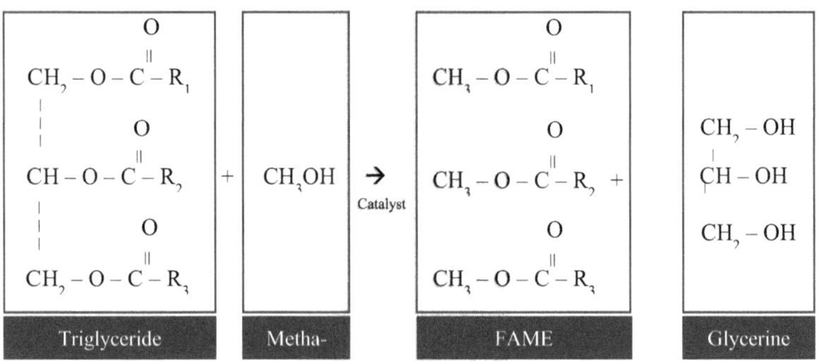

[136] Demirbas, 2008.
[137] Van Gerpen, 2005, Freedman, 1984.
[138] Van Gerpen, 2005.
[139] Van Gerpen and Knothe, 2005.
[140] Kaltschmidt, 2009.
[141] Own demonstration based on Knothe, 2005.

In contrast to petro diesel, biodiesel possess various advantages. Additionally to being technically competitive, biodiesel is directly derived from a renewable (domestic) resource so that there is less dependence on petroleum producing countries. Furthermore, the fuel is absolutely biodegradable and reduces therefore GHG emissions to which aspect Chapter 3.6 will present a more detailed overview. Moreover, a higher flash point allows a safer handling and storage of the fuel as well as better lubricity levels can improve the quality of your diesel blend by adding biodiesel.

3.5. Trade flows and concrete example of a Supply Chain Manager

In the following section current oilseed and vegetable oil as well as the biodiesel trade flows will be presented in order to provide an overall view of the interconnectedness of this global market. After outlining the principal flows of oilseeds and oilseeds products including meal and oils, a major focus will be laid on soybean-based biodiesel (soybean methyl ester – SME). On the origin side, Argentina and Brazil as well as the US will be relevant countries, while on the demand side; the major markets are China, Europe and India as demonstrated in Figure 22. Moreover, Figure 22 and Figure 23 demonstrate the main importing and exporting countries of selected oilseeds including soybeans (SB), rapeseed seeds (RSS) and sunflower seeds (SFS), whereas Figure 24 and Figure 26 summarize the vegetable oil and methyl ester flows referring to soybean- (SBO), sunflower- (SFO), rapeseed- (RSO) and palm oil (PO) as well as palm- (PME) and soybean methyl ester (SME).

As aforementioned, Brazil and Argentina together export over 80 per cent of worldwide SB flows while Argentina mainly processes the SB to SBO and SME and then operates as biggest exporter of SBO (58 per cent) with almost four mmt within a global vegetable oil market of approximately 52 mmt of trade flows per year. Besides SBO, Argentina also exports 100 per cent of the SME while over 80 per cent of PME is exported from Indonesia. In contrast to exporting countries, major importing ones are the EU27, India and China.[142]

[142] USDA, 2011h.

Figure 21: Global soybean trade flows scheme in 2010 (mmt) [143]

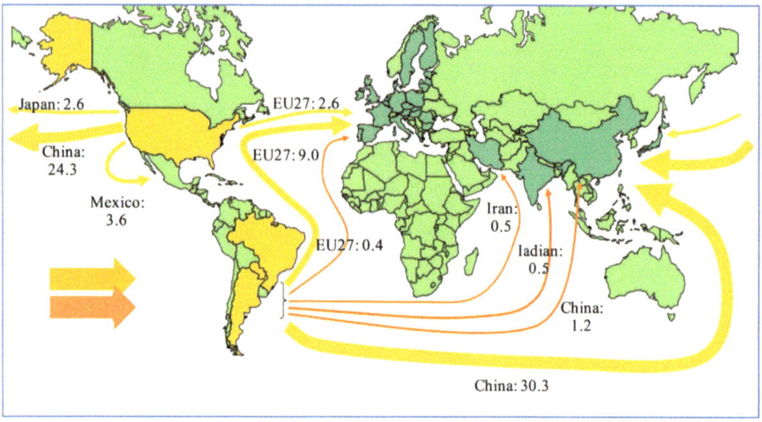

The EU27 imports basically all the major seeds and oils and additionally produces locally an increased amount of rapeseed, particularly in France and Germany. While the oils go directly to biodiesel production and the food industry, the seeds are destined to the 500-plus European crushing plants which amount to an overall capacity of over 100 mmt. At present, most of these crushing plants cannot compete with South American and Asian plants so that the current utilization rate is much lower as well as 40 mmt of oilseeds are processed to gain the meal and oil products. In contrast to the EU27, China basically imports almost two thirds (62 per cent) of global SB for the local crushing industry which developed a total capacity of approximately 120 mmt per year in order to produce SBO and SBM in order to mainly produce protein-rich animal feed for the Chinese livestock industry. Out of these 120 mmt, less than 50 per cent has been crushed (55 mmt of SB).[144] Therefore, the Chinese market currently remains with a large overcapacity that pushes the profit margins down reducing the overall crush activity and further investments. Thus, a considerable amount of SBs are directly transported to the Chinese State Reserve (CSR) and are kept as inventory. Given the massive volumes China purchases, the State Owned Enterprises (SOE) possess an enormous bargain power within the market and can literally influence prices to a certain degree. In addition to SB, the major exporters for SBO are Argentina and Brazil (Figure 26). Whereas both countries are major producers of SB, Brazil

[143] OVID 2012 based on AEE, 2009, GTIS, 2012, EUROSTAT, 2011, Oil World 2009 and OVID 2012.
[144] Noble Group Research, 2012b.

differs from Argentina in terms of export tax incentives as described in Chapter 4.1. Most of the SB are directly exported to Europe or China while Argentina's crushing and refining industry enjoys taxes incentives on SBO and SME, that is, it is more attractive to export the SB complex in the form of oil, methyl ester and meal, versus the initial seed. Figure 24 and Figure 26 show the vegetable oil and methyl ester flows.

Figure 22: Major exports of oilseeds[145]

(kmt)	2012/13	2011/12	2010/11	2009/10	2008/09	2007/08
SB		**94'867**	**88'775**	**91'135**	**78'850**	**79'270**
Argentina		8'480	8'545	12'635	6'955	14'455
Share		8.9%	9.6%	13.9%	8.8%	18.2%
Brazil		36'695	29'085	28'300	30'285	25'355
Share		*38.7%*	*32.8%*	*31.1%*	*38.4%*	*32.0%*
USA		39'145	40'875	40'830	34'840	32'570
Share		*41.3%*	*46.0%*	*44.8%*	*44.2%*	*41.1%*
RSS		**12'720**	**10'430**	**10'555**	**11'695**	**5'870**
Austtralia		2'170	1'470	1'240	975	465
Share		17.1%	14.1%	11.7%	8.3%	7.9%
Canada		9'050	7'110	7'145	7'700	4'465
Share		*71.1%*	*68.2%*	*67.7%*	*65.8%*	*76.1%*
SFS		**1'665**	**1'255**	**1'295**	**1'535**	**880**
EU27		775	490	620	350	485
Share		46.5%	39.0%	47.9%	22.8%	55.1%
Ukraine		475	445	360	770	70
Share		*28.5%*	*35.5%*	*27.8%*	*50.2%*	*8.0%*

As aforementioned, Brazil and Argentina combined export over 80 per cent of worldwide SB flows while Argentina mainly processes the SB to SBO and SME and then operates as biggest exporter of SBO (58 per cent) with almost four mmt within a global vegetable oil market of approximately 52 mmt of trade flows per year. Besides SBO, Argentina also exports 100 per cent of the SME while over 80 per cent of PME is exported from Indonesia. In contrast to exporting countries, major importing ones are the EU27, India and China. The EU27 imports basically all the major seeds and oils and additionally produces locally an increased amount of rapeseed, particularly in France and Germany. While the oils go directly to biodiesel production and the food industry, the seeds are destined to the over 500 European crushing plants which amount up to an overall capacity of over almost 100 mmt. At present, most of these crushing plants cannot compete with South American and Asian plants so that the current utilization rate is far below 50 per cent and only 40 mmt of oilseeds are processed to gain the meal and oil products.

[145] Own demonstration based on Oil World, 2009, Noble Group Research, 2012 and Industry Consensus.

In contrast to the EU27, China basically imports almost two thirds (62 per cent) of global SB for the local crushing industry which developed a total capacity of approximately 120 mmt in order to produce SBO and SBM in order to mainly produce protein-rich animal feed for the Chinese livestock industry. Out of these 120 mmt, less than 50 per cent is used as only 55 mmt of SB are crushed. Therefore, the Chinese market currently remains with a large overcapacity that pushes the profit margins down and reduces the overall crush activity.

Figure 23: Major imports of oilseeds[146]

(kmt)	2012/13	2011/12	2010/11	2009/10	2008/09	2007/08
SB		**94'867**	**88'775**	**91'135**	**78'850**	**79'270**
EU27		11'925	12'495	13'240	12'790	14'460
Share		*12.6%*	*14.1%*	*14.5%*	*16.2%*	*18.2%*
China		58'900	51'725	52'170	40'770	37'870
Share		*62.1%*	*58.3%*	*57.2%*	*51.7%*	*47.8%*
RSS		**12'720**	**10'430**	**10'555**	**11'695**	**5'870**
China		2'510	1'055	2'250	2'775	530
Share		*19.7%*	*10.1%*	*21.3%*	*23.7%*	*9.0%*
EU27		3'145	2'900	1'950	3'000	435
Share		*24.7%*	*27.8%*	*18.5%*	*25.7%*	*7.4%*
Japan		2'375	2'310	2'215	2'135	1'705
Share		*18.7%*	*22.1%*	*21.0%*	*18.3%*	*29.0%*
SFS		**1'665**	**1'255**	**1'295**	**1'535**	**880**
EU27		255	265	190	525	190
Share		*15.3%*	*21.1%*	*14.7%*	*34.2%*	*21.6%*
Turkey		820	580	670	410	496
Share		*49.2%*	*46.2%*	*51.7%*	*26.7%*	*56.4%*

Thus, a considerable amount of SBs are directly transported to the Chinese State Reserve (CSR) and are kept as inventory. Given the massive volumes China purchases, the State Owned Enterprises (SOE) possess an enormous bargain power within the market and can literally influence prices to a certain degree.[147] In addition to SB, the major exporters for SBO are Argentina and the United States as well as Brazil. Whereas, those countries are major producers of soybeans, for instance Brazil differs from Argentina in terms of export tax incentives. Most of the SB are directly exported to Europe or China while Argentina's crushing and refining industry enjoys from taxes on SBO and SME so that profit margins are much better for soybean and sunflower complex products, that is oil, methyl ester and meal, versus the initial seed. In Chapter 4, political incentives and regulations will be explained in more detail.

[146] Own demonstration based on Oil World, 2009, Noble Group Research, 2012 and Industry Consensus.
[147] Lui, 2012.

Figure 24: Major imports of vegetable oils and methyl esters[148]

(kmt)	2012/13	2011/12	2010/11	2009/10	2008/09	2007/08
Ttl oils		**50'850**	**48'960**	**46'790**	**46'810**	**44'980**
China		**8'160**	**7'620**	**8'195**	**9'015**	**9'105**
Share		*16.0%*	*15.6%*	*17.5%*	*19.3%*	*20.2%*
India		**9'360**	**8'410**	**8'765**	**8'350**	**6'075**
Share		*18.4%*	*17.2%*	*18.7%*	*17.8%*	*13.3%*
EU27		**6'585**	**7'015**	**7'425**	**7'595**	**7'590**
Share		*12.9%*	*14.3%*	*15.9%*	*16.2%*	*16.9%*
SBO		**6'410**	**8'415**	**8'070**	**8'240**	**10'375**
China		1'300	1'285	1'523	2'523	3'005
Share		*20.3%*	*15.3%*	*18.9%*	*30.6%*	*29.0%*
India		950	1'080	1'655	925	610
Share		*14.8%*	*12.8%*	*20.5%*	*11.2%*	*5.9%*
EU27		305	775	515	585	1'075
Share		*4.8%*	*9.2%*	*6.4%*	*7.1%*	*10.4%*
RSO		**3'155**	**3'310**	**2'715**	**2'240**	**1'850**
China		665	680	780	435	250
Share		*21.1%*	*20.5%*	*28.7%*	*19.4%*	*13.5%*
EU27		5'535	565	345	375	255
Share		*175.4%*	*17.1%*	*12.7%*	*16.7%*	*13.8%*
USA		1'430	1'405	1'050	1'025	940
Share		*45.3%*	*42.4%*	*38.7%*	*45.8%*	*50.8%*
PO		**35'950**	**33'560**	**32'405**	**32'360**	**29'890**
China		6'005	5'635	5'755	5'945	5'840
Share		*16.7%*	*16.8%*	*17.8%*	*18.4%*	*19.5%*
EU27		4'955	4'900	5'720	5'590	4'970
Share		*13.8%*	*14.6%*	*17.7%*	*17.3%*	*16.6%*
India		7'305	6'575	6'535	6'800	5'440
Share		*20.3%*	*19.6%*	*20.2%*	*21.0%*	*18.2%*
SFO		**5'345**	**3'675**	**3'605**	**3'975**	**2'865**
EU27		790	780	840	1'050	1'290
Share		*14.8%*	*21.2%*	*23.3%*	*26.4%*	*45.0%*
India		1'090	755	570	625	25
Share		*20.4%*	*20.5%*	*15.8%*	*15.7%*	*0.9%*
Turkey		425	365	225	335	230
Share		*8.0%*	*9.9%*	*6.2%*	*8.4%*	*8.0%*
PME		**1'945**	**1'210**	**550**	**425**	**610**
Canada		250	-	-	-	-
Share		*12.9%*	*0.0%*	*0.0%*	*0.0%*	*0.0%*
EU27		1'680	1'175	445	275	185
Share		*86.4%*	*97.1%*	*80.9%*	*64.7%*	*30.3%*
SME		**2'070**	**1'530**	**1'375**	**910**	**345**
EU27		1'905	1'385	1'290	715	25
Share		*92.0%*	*90.5%*	*93.8%*	*78.6%*	*7.2%*

In addition to SB, PO does not have the products and needs to be crushed immediately after harvesting. As the major producers are Malaysia and Indonesia, both basically drive the global PO trade flows mainly into India, China and Europe. The PO is comparatively cheap but has a generally lower quality than Rape or Soybean oil so that it is not an ideal substitute for the others.[149] Additionally, FAME need to comply with certain quality standards and among others possess a

[148] Own demonstration based on Oil World, 2009, Noble Group Research, 2012 and Industry Consensus.
[149] Simantob, 2011

cold filter plugging point (CFPP) that provides you an indication about the temperature at which the biodiesel starts to generate gums causing troubles for motor engines.[150] So, depending on the blend and portions of RME, SME and PME or tallow, this CFPP can vary between -10° Celsius and +15 so that it might become critical particularly during winter time which oil refineries use. The CFPP can also be lowered with chemical additives (e.g. mixtures of ethylene vinylacetate copolymers and naphthenic distillates) but it is not a common practice as distributors like Shell and BP do not appreciate an excessive use.[151] These technical limitations restricted a certain amount of SBO and PO within the blend, especially in northern Europe where FAME with lower CFPP (approximately -10° till -20° Celsius) is necessary.[152] In spite of the aforementioned small share in global vegetable oil trade flows and thanks to the enhanced quality, RSO bear a significant role within the EU27 as indicated in Figure 25.

Figure 25: Average feedstock use in major European countries in 2011[153]

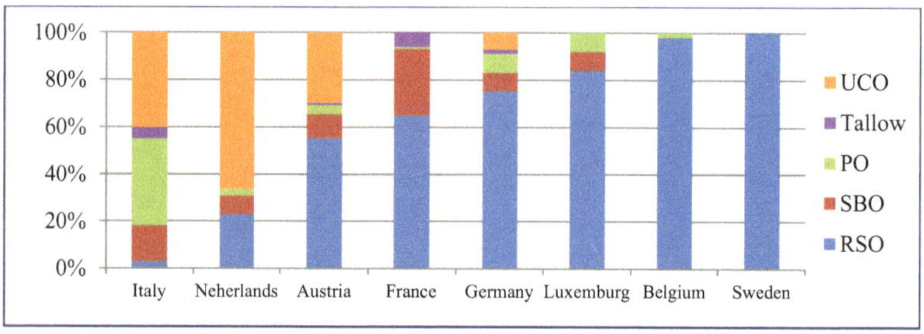

Finally, Ukraine and Russia are two countries with an enormous sunflower production, which is mainly destined to export. While Ukraine exports both, seeds and oil, Russia mainly crushes the seeds locally and then exports the oil.[154] Major destinations are the EU27 and India as well as Turkey, which serves as a hub for further distribution, e.g. to Iran, Pakistan as well as Egypt. Egypt, for example, is a country that substituted entirely its SBO imports from South America with the comparatively cheaper sunflower oil from Ukraine and Russia. Approximately

---

150 Vasquez, 2012.
151 Zylla, 2012.
152 Kabath, 2011.
153 Greenpeace, 2011.
154 Soroko, M., 2012

eighty per cent of palm oil is produced in Malaysia and Indonesia so that the major flows come from South East Asia to China and Europe as well as India.[155]

Figure 26: Major exports of vegetable oils and methyl esters[156]

(kmt)	2012/13	2011/12	2010/11	2009/10	2008/09	2007/08
Ttl oils		**50'850**	**48'960**	**46'790**	**46'810**	**44'980**
SBO		**6'410**	**8'415**	**8'070**	**8'240**	**10'375**
Argentina		3'805	4'560	4'395	4'765	6'045
Share		*59.4%*	*54.2%*	*54.5%*	*57.8%*	*58.3%*
Brazil		1'495	1'575	1'415	1'665	2'365
Share		*23.3%*	*18.7%*	*17.5%*	*20.2%*	*22.8%*
USA		505	1'465	1'525	995	1'320
Share		*7.9%*	*17.4%*	*18.9%*	*12.1%*	*12.7%*
RSO		**3'155**	**3'310**	**2'715**	**2'240**	**1'850**
Canada		2'405	2'455	1'920	1'570	1'330
Share		*76.2%*	*74.2%*	*70.7%*	*70.1%*	*71.9%*
PO		**35'950**	**33'560**	**32'405**	**32'360**	**29'890**
Indonesia		17'340	15'500	14'990	15'540	13'720
Share		*48.2%*	*46.2%*	*46.3%*	*48.0%*	*45.9%*
Malaysia		17'885	17'150	16'610	15'990	15'040
Share		*49.7%*	*51.1%*	*51.3%*	*49.4%*	*50.3%*
SFO		**5'345**	**3'675**	**3'605**	**3'975**	**2'865**
Argentina		880	660	570	810	1'085
Share		*16.5%*	*18.0%*	*15.8%*	*20.4%*	*37.9%*
Russia		1'115	195	505	835	325
Share		*20.9%*	*5.3%*	*14.0%*	*21.0%*	*11.3%*
Ukraine		3'205	2'650	2'380	2'210	1'345
Share		*60.0%*	*72.1%*	*66.0%*	*55.6%*	*46.9%*
PME		**1'945**	**1'210**	**550**	**425**	**610**
Indonesia		1'575	1'180	425	185	455
Share		*81.0%*	*97.5%*	*77.3%*	*43.5%*	*74.6%*
Malaysia		370	25	120	240	150
Share		*19.0%*	*2.1%*	*21.8%*	*56.5%*	*24.6%*
SME		**2'070**	**1'530**	**1'375**	**910**	**345**
Argentina		2'070	1'530	1'375	910	345
Share		*100.0%*	*100.0%*	*100.0%*	*100.0%*	*100.0%*

Referring to the EU 27, Figure 27 represents the crush and trade activity of major selected oilseeds in volume terms. EU27 intra trade is excluded from the import-export flows given that within the Schengen Area small volume trade flows are difficult to track. In general Soybean import, export and crushing have remained stable in the recent years. Whilst dependent on the price spreads between the products and the substitutability of both seeds for crush and oils for refineries, the overall amount of vegetable oil and oilseeds has been stable during the last years. In the end, this table demonstrates a consolidated picture.[157] Nevertheless, in reality there are various market players that are specialized or even positioned within a niche so that for example they focus exclusively on Rape, Palm or even

[155] Noble Group Research, 2012b.
[156] Own demonstration based on Oil World, 2009, Noble Group Research, 2012 and Industry Consensus.
[157] Simantob, 2011.

used cooking oils accordingly. Depending on upcoming regulations, particularly sustainability certifications and double counting schemes for lower GHG emissions of individual feedstock, existing preferences might shift and create new competitive advantages and creates business opportunities for current as well as future market participants.

Figure 27: EU 27 consolidated trade flow data[158]

(kmt)		2012/13	2011/12	2010/11	2009/10	2008/09	2007/08
SB	Crush		11'900	12'450	12'650	12'957	14'912
	IM		11'770	12'798	13'509	13'331	15'215
	EX		40	40	40	40	40
SF	Crush		6'800	6'000	5'950	5'920	4'280
	IM		295	325	295	595	299
	EX		760	492	660	305	495
RS	Crush		21'700	22'450	22'987	20'430	18'450
	IM		3'330	3'070	2'160	3'340	687
	EX		95	105	190	87	55
Oil	**Ttl production**		**14'217**	**14'032**	**14'507**	**13'710**	**12'381**
	Ttl IM		**6'973**	**7'667**	**7'896**	**8'108**	**7'542**
	Ttl supply		**22'402**	**22'741**	**23'559**	**22'778**	**20'804**
	Ttl demand		**20'614**	**20'526**	**21'715**	**20'822**	**19'074**
	Ttl biodiesel		**3'587**	**2'556**	**1'732**	**991**	**209**
	SME IM		1'904	1'383	1'288	719	24
	PME IM		1'683	1'173	444	272	185

Figure 28 shows a concrete example of the global soybean trade flows of one of the major agriculture supply chain manager. Noble Group is a Supply Chain Manager and reaches a global scale with its operations encompassing a total of 150 offices in 38 countries.

Figure 28: Soybean trade flows from a selected supply chain managing company[159]

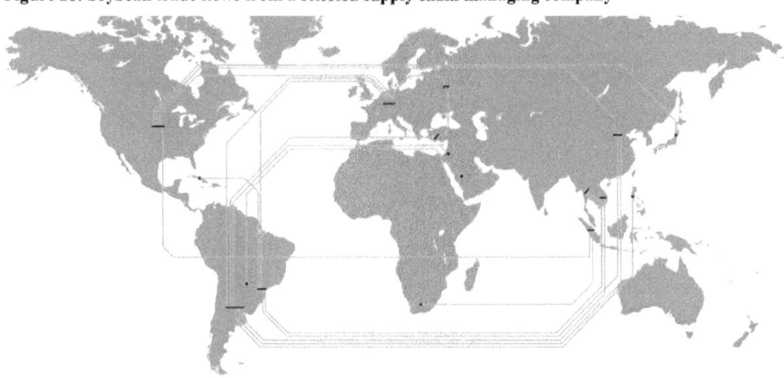

[158] Own demonstration based on Oil World, 2009, Noble Group Research, 2012 and Industry Consensus.
[159] Own demonstration based on Noble Group's website information an internal data.

The Fortune 500 company is based in Hong Kong (China), listed in Singapore and operates regional headquarters in Geneva (Switzerland), London (UK) Stamford (USA), Sao Paulo (Brazil) and Buenos Aires (Argentina). Currently the company employs around 12,000 employees and generates an annual turnover of US$60 billion with a net profit of approximately US$500 million whereas the agriculture products segments represent approximately one third of their business activity besides other sectors such as energy and logistics.[160]

In 2011 the total biodiesel import to EU27 rose above 2.5 mmt representing around 20 per cent of the local biodiesel production. The main origins were the Argentina and Indonesia with SME and PME respectively. (Figure 27 and Figure 29). As initially outlined, the EU27 will create a future demand of 25mmt of biodiesel and 15mmt of ethanol while only producing about 21mmt and 6mmt accordingly by reaching an energy share of 10 per cent of biofuels in total fuel consumption in 2020 (Figure 1).

Figure 29: EU27 biodiesel imports 2010[161]

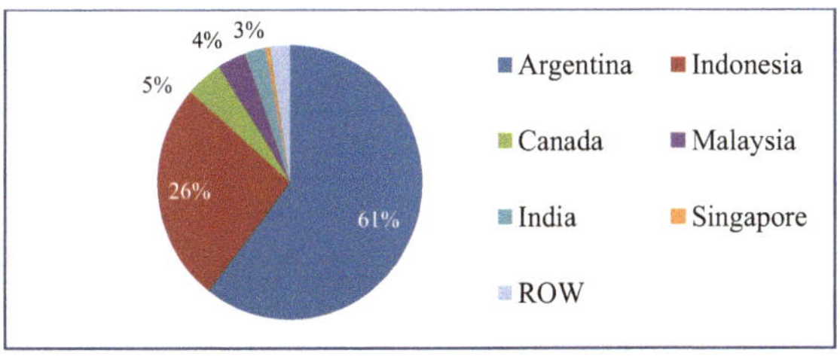

The supply-demand gap will be adjusted via imports mainly from Argentina and Indonesia in the case of biodiesel (each 2mmt) and from Brazil regarding ethanol. In general, there are comparatively low volumes of international trade in biofuels at present and most of these volumes are limited to the major feedstock in terms of oilseeds and vegetable oil. Moreover, most of the current biofuels are being produced and consumed in domestic markets. Though, considering political frameworks and regulations for blending requirements securing future demand among others, volatile oil prices, stable and profitably margins attracting agricultural trading houses trade in biofuels will altogether be likely to increase interna-

160 Noble Group Corporate presentation Slides will be presented in the Annex
161 EUROSTAT, 2012.

tional trade.[162] So that as aforementioned, biofuel production will quadruple the next coming twenty years up to as much as ten per cent of global fuel consumption.[163]

In general, in 2011 the Organisation for Economic Co-operation and Development countries (OECD) consumes approximately 47 million oil barrels a day (mb/d) whereas the global oil demand is expected to rise up to 90 mb/d in 2012.[164] While the Aggregated Demand (AD) from Northern countries rises, they do not have the land availability to meet the energy needs. Not only does land availability inhibit any large-scale biofuels production in the North, there are additional factors such as climate conditions and higher costs. The best circumstances for most energy efficient feedstock production, that is, SBO and PO for biodiesel and sugar cane for bioethanol, are in Southern hemisphere countries with tropical and sub-tropical climates. Despite the lack of infrastructure in a wider and narrower sense as well as developed logistics, suitable soils, longer growing seasons and lower labour costs are available. Thus, it is a more cost-efficient and energy efficient approach to produce biofuels in the South. The geographical asymmetry between demand and supply, North and South, has so far been insignificant as there are presently low volumes of international trade. An estimation is that sugar cane production could respond to 10 per cent of the gasoline use globally, which means a possibility for Southern countries to become producers and exporters of a value added product much in demand.[165] Currently, there are a number of obstacles on an international level that impede any greater volume of biofuel trade, those generated within the international trading system itself due to a lack of clear classification by the WTO (World Trade Organization) and in addition to that those posed by national protective measures, such as tariffs, tariffs escalation, quotas, technical, environmental and social standards.[166] Generally, energy trade has not been of utmost importance within the WTO. One of the reasons is that few of the main oil exporting countries, for example Saudi Arabia, have not been WTO members and dealt issues within the OPEC. There are also certain exceptions in the General Agreement on Tariffs and Trade (GATT), which is the main

---

[162] IPC, 2006.
[163] IEA, 2011.
[164] IEA, 2012.
[165] Worldwatch Institute, 2007.
[166] Dufey, 2006.

agreement of the WTO for trade in goods, which have left energy policies outside the discipline. Currently ethanol is classified as an agricultural product, whereas biodiesel is classified as an industrial product. The difference is very important because the rules of the trading system related to market access and subsidies generally treat agricultural products differently from industrial products.[167]

In addition to these classification issues, it is also the possibility of having biofuels classified as environmental goods. Thus, this would make biofuels subject to special negotiations to reduce tariffs barriers according to paragraph 31 of the Doha Ministerial Declaration,[168] a paragraph that calls for "the reduction or, as appropriate, elimination of tariffs and non-tariff barriers to environmental goods and services".[169] Talks about the application of this paragraph and the development of criteria to include an environmental good had not progressed far when the Doha Round was suspended. Brazil and India wanted negotiations on Environmental Services and Goods (ESG) on biofuels,[170] and so did the European Community (EC). Renewable energy technology has been discussed as having the potential to be classified ESG.[171] There are three kinds of possible environmental goods that have been discussed. Low carbon-fuels, the category of ethanol or biodiesel, renewable technologies as solar cells or wind turbines, or energy efficient environmentally preferable products (EPPS), as for example more efficient refrigerators.[172] The definition for ESG has to be agreed upon by all WTO members though, and it has not been agreed upon that environmental goods include renewable energy products or biofuels. Depending on environmental performance considering life cycle assessments (LCA) results, there might be a shift within the next decade.

On the one side, a more conservative scenario given by Beyond Petrol (BP) in 2011 estimated the biofuel share at three per cent in 2011 with an increase up to nine per cent by 2030 as in Figure 30. Earlier in 2005, the IEA projected the world

167 Vasquez, 2012.
168 WTO, 2001.
169 WTO, 2001.
170 Worldwatch Institute, 2007.
171 IPC, 2006
172 GRAIN, 2007c.

biofuel share only to six per cent of world transport fuels as it represented only 1.5 per cent seven years ago.[173]

Figure 30: Energy use in transport sector (million tons of oil equivalent)[174]

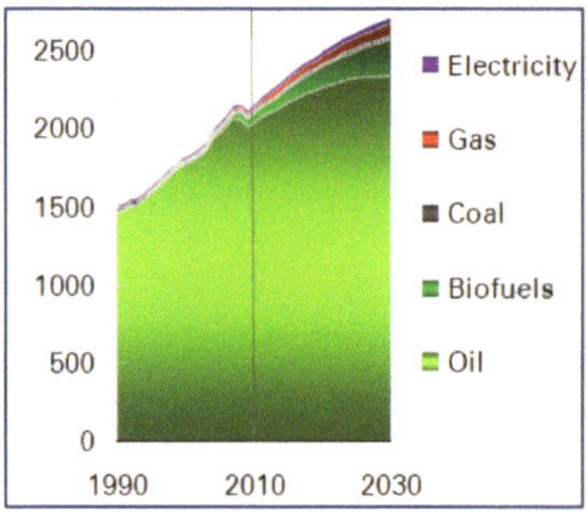

Nevertheless, in 2008, reports from the IEA have increased the figures, up to 10 per cent of transport fuels by 2025[175] and in the most recently published report, the IEA outlined a 2050 scenario, in which biofuels could supply up to 27 per cent of total transport fuel and contribute in particular to the replacement of diesel, kerosene and jet fuel as in Figure 31. The estimated biofuels use could therefore avoid approximately 2 billion tons of CO2 emissions yearly in case of sustainable production.[176] In order to reach this ambitious goal, conventional biofuel technologies need to enhance in terms of efficiency, cost and general sustainability criteria.

Figure 31: Global Energy use in transport and biofuel use within the transport sector[177]

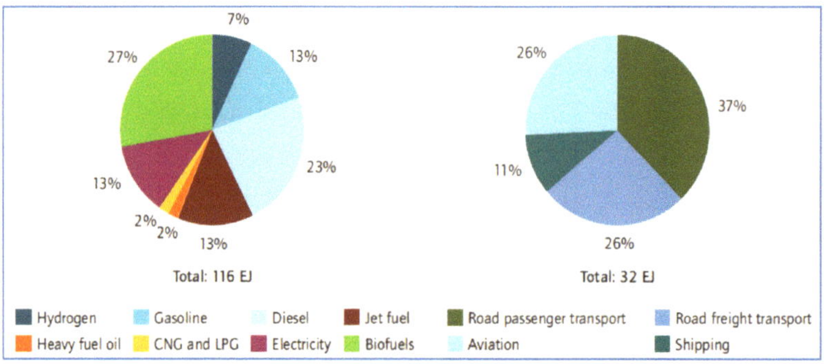

Additionally, second generation biofuels would need to be commercially deployed, requiring significant investments in research and development. The EU27 RED target of 10 per cent of biofuel blends within the transport sector by

[173] IEA, 2004.
[174] BP, 2011.
[175] IEA, 2008.
[176] IEA, 2011.
[177] IEA, 2010. CNG stands for compressed natural gas and LPG for liquefied petroleum gas.

2020contributes to this vision.[178] Nonetheless, in order to achieve the political target, the EU27 need to import up to 50 per cent of biofuels outside the Schengen area.[179] Some countries are already preparing to meet the increasing demand, as Malaysia and Indonesia that have expanded the palm oil plantations substantially. While Brazil and Argentina are expected to further expand the export sector for SME, Asian countries including Japan, Korea and Taiwan are expected to become more relevant import markets.[180]

3.6. GHG emissions as main sustainability criteria for biodiesel

Analysing the environmental aspects of biodiesel, it is essential to consider two main environmental impacts of the biofuel industry. First, the major reason for worldwide biofuels promotion is environmental quality, that is, the ability to reduce GHG emissions in comparison with regular diesel fuels. Second, externalities generated by biofuel production throughout the entire life cycle are likewise relevant for a sound environmental impact analysis including subsequent consequences from the overall effect of the production, its affects on the ecosystem, the expansion of the agricultural frontier and deforestation among others. [181]

The RED encourages particularly the use of alternatives to fossil fuels that go beyond first generation biofuels. Thus, for instance next generation biofuels count double and cars on green electricity count twice and a half within the target calculations, whereas first generation biofuels are only count once. The GHG emission savings from the use of biofuels have to be at least 35 per cent. If the biofuel plant was already in operation before January 23rd 2008 it has to comply with this requirement by 2013. If the plant came into operation after that date it is already required to obtain the 35 per cent savings as the RED has been transposed into national legislation. Then, plants have to comply with 50 per cent GHG emission savings requirements and for all post-2017 plants the target is 60 per cent by 2017. The RED set sustainability criteria for biofuels irrespective of whether the raw materials were produced inside or outside the territory of the EU.

[178] EREC, 2008.
[179] IEA, 2008.
[180] IEA, 2011.
[181] Asal, 2006.

Thus, biofuel plants and feedstock producers, both have to keep in mind these sustainability criteria.

Climate Change is the overall threat whereas biofuels are considered as only (a small) part of an integrated solution. Within the biofuels sphere, the GHG emissions for SME should be evaluated on a life cycle assessment (LCA) or well-to-wheel (WTW) analysis basis. A WTW assessment can be broken down into well-to-pump (WTP) and pump-to-wheels (PTW) stages, whereas the former commences with the feedstock recovery, followed by biodiesel production, and terminates with the availability at the pump, while the latter represents the distribution and operation activities either passively or actively, that is, the vehicle's operating activity.[182] Thus, when evaluating the energy and GHG emission implications of biodiesel as well as advanced technologies of production and vehicles, a WTW or "cradle-to-grave" (CTG) assessment provide essential insights.[183] So, the total LCA GHG emissions encompass all net emissions of CO_2, CH_4 and N_2O that can be assigned to the biodiesel (or any blend. Moreover, comprising all relevant stages from extraction or cultivation, involving land-use changes, transport as well as distribution, processing and combustion, irrespective of where those emissions happen.[184]

Despite various independent and reliable LCAs, which result in a positive accounting, that is a favourable GHG balance for first generation biodiesel compared to traditional diesel, they differ in terms of the actual amount of GHG emission savings depending on numerous factors such as chosen feedstock composition or refinery technologies. The relevant studies cited by the IEA all show a positive GHG balance for biodiesel.[185] Based on numbers elaborated by the German Agency for RE, Figure 32 provides a brief overview about the CO2 emission abatement per ha RS and corn accordingly. Furthermore, numerous environmental and social standards during their production have to be assessed and certification criteria established. The most important criteria upon which liquid biofuels and their feedstock options as well as the production process

[182] Teulon, 2011 and Elgowainy, 2009.
[183] Elgowainy, 2009.
[184] European Parliament and Council, 2009.
[185] IEA, 2004.

should be evaluated, in form of a Life Cycle Assessment (LCA) will be outlined in the following pages.

Figure 32: CO2 emission abatement per ha rapeseed and corn through RME and Bioethanol (mt/ha)[186]

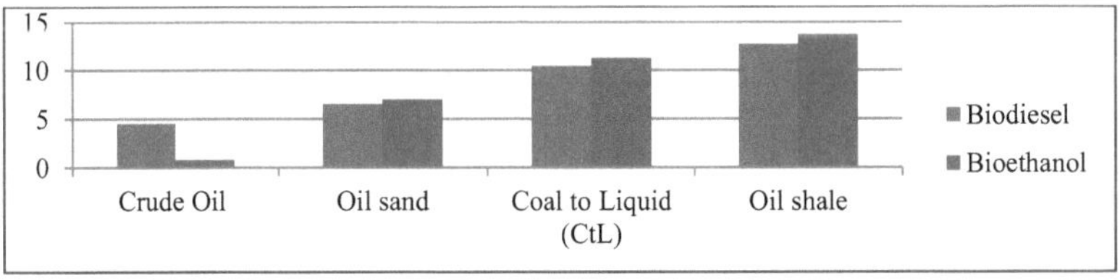

In 1998, a first comprehensive life-cycle assessment (LCA) for SME was carried out by the National Renewable Energy Laboratory (NREL), and the energy inventory for this analysis was updated in 2009 applying 2002 data. [187], The continual adoption of new technologies in SB cultivation, crushing and transesterification has been affecting the life-cycle energy use over time, requiring that LCA need to be updated as often as possible. Despite a variety of renewability measurements, the Fossil Energy Ratio (FER) has been used in numerous LCA so that a representative picture of sustainability can be obtained and particularly to evaluate fuel renewability. FER is defined as: FER equals the Renewable fuel energy output divided by the biodiesel share of fossil energy input. [188] Only fossil (non-renewable) energy is included in the denominator and the ratio does not consider renewable energy sources like solar and wind. Nevertheless, the FER does not measure system efficiency, as fossil fuel can be replaced by other renewable fuel. Thus, the net energy ratio (NER), taking into consideration the total energy input can be applied to measure this efficiency, rather than FER.[189] Whereas an increased FER represents higher renewability of biodiesel, it does not guarantee that biodiesel becomes higher economic viability. The present analysis shows that the FER of SME was 5.5 based on agricultural data input of 2006 versus 3.2 applying the 1998 dataset. The improvements are mainly thanks to increased SB yields[190] and higher energy-efficient crushing and transesterification plants.[191] The energy input in SB farming was decreased by over 50 per cent, in

[186] Own demonstration based on AEE, 2009.
[187] Sheehan, 1998.
[188] Pradhan, 2011.
[189] Pradhan, 2008.
[190] Higher production yields are primarily driven by GMO improvements, Monsanto, 2012 and Editora Gazeta, 2011.
[191] Vasquez, 2012.

crushing by almost 60 per cent and in transesterification by one approximately 30 per cent per ton of produced methyl ester.[192] So that the overall energy input could be lowered by around 40 per cent for the same quantity of biodiesel. The addition of indirect energy uses through secondary inputs, including farm machinery and other materials has a minor effect on the FER. Moreover, the FER of SME tends to keep increasing due to improving yields and better infrastructure in terms of logistics and energy-efficient technologies.[193]

In general, the LCA was divided into four subsystems: feedstock production, feedstock transportation, crushing and transesterification and finally biodiesel distribution. Therefore, an inventory of material and energy was established quantifying non-RE inputs in each subsystem. Both, direct and indirect energy sources were included including liquid fuel and electricity used to directly power equipment. Furthermore, energy content of additional materials, such as fertilizers, pesticides, and other petrochemicals had been taken into consideration as illustrated in Figure 33. The energy in form of labour was disregarded so that in the inventory list was converted to their equivalent life-cycle energy content. Life Cycle Inventory (LCI) is the accounting of direct and indirect inputs as well as process outputs during the entire life cycle of a product. The LCI for SME in this particular case considers the four aforementioned subsystems. Feedstock production, energy transport and logistics of soybeans to the crush and refinery facilities, energy for the crushing and transesterification processes themselves. The crush (into oil and meal) can be executed either with mechanical extruders or more commonly via chemical extraction using hexane.[194] The transesterification as aforementioned mainly operates with methanol and sodium hydroxide or sodium methylate as catalysts in reactors.

As seen in the chapter about commodity trade flows, soybean demand is driven by the demand for the meal used as a high protein in the animal feed industry.[195] Historically crush margin has been valuing meal comparatively higher than the oil.[196] In comparison to RME production in which the oils is the premium

---

[192] Pradhan, 2009.
[193] Vasquez, 2012.
[194] Motenegro, 2011 and Noble Argentina, 2011.
[195] World Bank, 2008b.
[196] Gallo, 2011.

product and the Rapeseed meal (RSM) is considered as a residual, the SBM is not at all a secondary product of SME production. Both soy complex products are jointly produced and sold in separate markets as illustrated in Chapter 2 below the SB and SB products production flow (Figure 13).

Figure 33: Life Cycle energy equivalents

Input	Embedded energy	Life-Cycle energy (%)	Life-Cycle energy equivalent
Fuel input			
Diesel	36 MJ/l[197]	84[198]	43 MJ/l
Gasoline	32 MJ/l[197]	81[198]	41 MJ/l
Propane	24 MJ/l[197]	90[198]	26 MJ/l
Natiral gas	37 MJ/m3[197]	94[198]	39 MJ/m3
Electricity	4 MJ/kWh	33[199]	7 MJ/kWh
Steam at 10 bar	2 MJ/kg	61	3 MJ/kg
Material inputs			
Nitrogen	-	-	52 MJ/kg[200]
Phosporus	-	-	9 MJ/kg[200]
Potassium	-	-	6 MJ/kg[200]
Herbicide	-	-	319 MJ/kg[200]
Inserticide	-	-	325 MJ/k[200]g
Lime	-	-	0.1 MJ/kg[201]
Seeds	-	-	5 MJ/kg[202]
Methanol	23 MJ/kg[203]	68[204]	34 MJ/kg[202]
Sodium methixide	-	-	32 MJ/kg[202]
Sodium hydroxide	-	-	2 MJ/kg[202]
Hydrochloric acid	-	-	2 MJ/kg[202]
Hexane	-	-	0.5 MJ/kg oil[197]

Therefore, all the energy used to produce SBM as a portion of the crush and the glycerine produced during transesterification have to subtracted from the SME LCI. Figure 35 provides a good energy use overview and leads to the final FER calculation. For the secondary product allocation, a mass based allocation method was used to determine how the energy consumption is attributed (Figure 34). Finally, the combination of the four subsystems produces the consolidated energy balance and the base case LCA for SME (Figure 35). After adjusting the inputs and energy allocating by secondary products, the total energy required to produce a litre of SME is 5.9 MJ. Transesterification requires the most energy, accounting for 56 per cent of the total energy consumed during the LCI while crushing requires approximately 20 per cent, followed by farming that accounts for

[197] Huo, 2008.
[198] Shapouri, 2002.
[199] EIA, 2010.
[200] Hill, 2006.
[201] Graboski, 2002.
[202] Sheehan, 1998.
[203] MI, 2009.
[204] Wang and Huang, 1999.

approximately 17 per cent. Thus, the calculated SME FER amounts to 5.5. This output has been increasing due to recent improvements in crushing and refining capacities. Higher volumes and enhanced technology facilitates energy efficient process so that the overall goal of reducing GHG and increasing the FER succeeds.[205].

Figure 34 & Figure 35: Base case energy use for SME and FER and mass based energy allocation

	Fossil energy use (MJ/l of SME)	
Subsystems	**Total used**	**Biodiesel fraction**
Beans cultivation	6.0	1.0
Beans transport	1.2	0.2
Crushing	6.3	1.1
Transesterification	4.0	3.3
SME transport	0.3	0.3
Total	17.8	5.9
SME output		32.7
FER		5.5

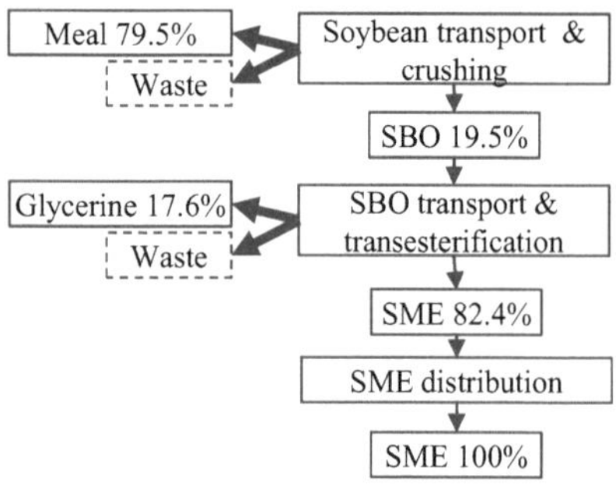

In contrast to the LCA, biofuels have specific GHG emission standard values summing up all of GHG emissions occurred in their life cycle. The Commission provides these data of different biofuel production pathways in the RED Annex.[206] Similar to the four stages during the LCA, the default GHG emission value three phases of the biofuel supply chain, such as cultivation, processing, and transport. The cultivation phase includes the emissions from the extraction or cultivation process itself; from the collection of raw materials; from waste and leakages; and the production of chemical or products use in extraction or cultivation. The processing phase includes emissions from the processing itself; from waste and leakages; and from the production of chemicals or products used in processing. The transport and distribution phase finally includes storage of raw materials and semi-finished materials and from the storage and distribution of finished materials.[207] As Figure 36 indicated, in case of RME, for instance, most of the GHG emission (29 gCO2eq/MJ) occur during the cultivation of raw material. It means even if the production process is improved, the high emissions are generated

[205] EPA. 2004.
[206] European Parliament and Council, 2009.
[207] Zylla, 2012 and Noble Argentina, 2011.

along the agricultural cultivation.[208] Between the feedstock there are major differences as for example, RS farming causes 11 gCO2eq/MJ more GHG emissions than during sunflower one. Part of the discrepancy can be derived from the utilization of fertilizer and therefore the indirect energy demand.

Figure 36: Standard values for GHG emissions[209]

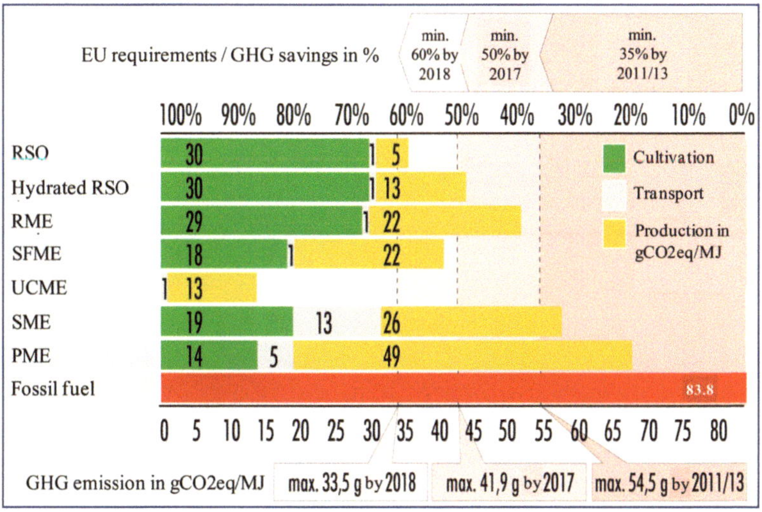

So, PME reaches the highest GHG emissions with 54 gCO2eq/MJ; nonetheless, PO still bears currently an enormous potential in changing the production process. A methane capture system at the palm oil mill could for instance significantly reduce GHG, down to 31 gCO2eq/MJ (35 per cent CO2equ/MJ savings). Moreover, GHG emissions that occur during the SME processing steps are higher than the rapeseed and sunflower ones. The major difference can be explained by the oil extraction process since the refining and transesterification process have similar GHG emission values. The LCI overview in Figure 34 confirmed already the intensive energy use via hexane to extract the SBO during the crush. In case of biodiesel produced with a feedstock, which has its base in used cooking oil (UCO), waste vegetable or animal oil the GHG emissions only account for 14 gCO2eq/MJ in total, consequently these oils can provide an attractive feedstock

[208] Montenegro, 2011.

[209] FNR 2011b based on RED 2009/28/EC values, future biofuels options, basis: estimated standard values based on 2009/28/EG.

for transesterification in terms of emission savings.[210] As the LCA confirmed, depending on every individual case, present GHG emission values can differ from default ones. Thus, actual GHG benefits could be more or less in line with those published in the RED Annex.

There are various other constraints which limit the comparability of LCA results and need to be considered when interpreting the results. For instance, results of life-cycle GHG balances may critically depend on the way land conversion related impacts are attributed. For instance, when oil palm plantations are established on converted natural forests and the associated emissions are depreciated over 100 years, GHG savings may result per hectare and year. Additional emissions will result if a depreciation period of 30 years is applied. When plantations are grown on tropical fallow (abandoned land), in general beneficial values result. Improvement of the product chain oriented life-cycle approach seems necessary, and is ongoing, but basic deficiencies may be overcome only through the use of complementary analytical approaches which capture the overall impacts of biofuels in the spatial and socio-economic context. This is necessary in particular to account for the indirect effects of land use change induced by increased demand.

Though, there are quite significant variations on GHG reductions, studies shows a number of different results, from negative to more than 100 per cent. It depends on feedstock, agricultural production, and conversion technology. Generally air emissions of key toxic substances are lower from biodiesel compared to conventional diesel over the life cycle. LCA studies are used as an indication of the energy needed to produce one unit of biodiesel, whether it takes more or less energy to produce it than the energy that is generated.[211] When estimating the energy balance one have to consider the total fuel cycle of the production, from the feedstock production to final consumption, the "well-to-wheel approach"[212] approach. There is quite a difference in results depending on origin, type of feedstock, agricultural methods and kind of conversion technology. Results may also differ as different methodologies are used when calculating the

[210] Greenea, 2011 and Mousdale, 2008.
[211] Dufey, 2006
[212] Elgowainy, 2009.

energy balance, for example depending on the assumptions made for the co-products. For instance, the bioethanol energy balance in Brazil is in a range between 3.7 till 10.2, averaging 8.3 units providing eight times more energy than the one unit necessary for its production. Studies from the EU on biodiesel made on rapeseed gives an energy balance between 0.33 and 0.82 units. The differences in energy balances depend on differences in the production process. As Brazil has good soil fertility, no extra inputs as fertilizers are needed, nor does it take much irrigation as the rainfall is good. In sugar cane production one also uses the bagasse-what is left of the crushed cane to generate energy to the production. So, the energy demand of fossil fuels is none. In the States the use of fertilizers and pesticides is much higher and the corn processing takes extra fossil fuels.

3.7. Criticism on criteria regarding GHG emissions and biodiversity

The actual GHG emission of every individual actor along the biodiesel life cycle results in a significant administrative burden for the whole supply chain. Therefore, the Commission provides a value for the average emissions from the fossil part of petrol and diesel consumed in the EU, which is 84 gCO2/MJ. Calculating the default GHG emission savings of different biofuel production pathways, the so-called fossil fuel comparator is used. The default GHG emission savings could be calculated by the following formula: GHG Savings = (Ef - Eb) /Ef, being Ef the total emissions from the fossil fuel comparator; and Eb the total emissions from the biodiesel. The Commission set these default GHG emission values based on its expanded analysis however producers may prove that their actual performance is better than these values. As aforementioned, particularly if a crusher or refiner applied more recent technologies or benefits from economies of scale so that energy efficiency increases. As in Figure 37, the spectrum on GHG is quite broad for different technologies as for instance with PME reaching from only 25 per cent up to 75 per cent depending on technology and extraction efficiency.

Figure 37: Life Cycle GHG balance of advanced and first generation biofuels[213]

[213]IEA, 2011, IEA, 2009, and OECD, 2008. Bio-SG is bio-synthetic gas, HVO stands for hydro-treated vegetable oil. The assessment exclude GHG emission from land use change. GHG savings above 100 per cent are feasible thanks to co-product generation.

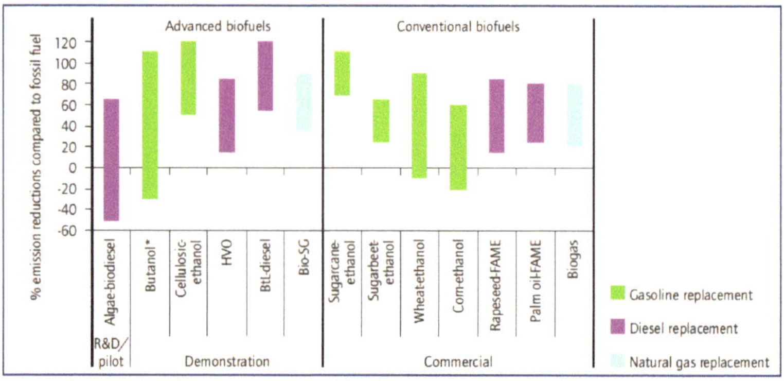

In addition to the GHG emissions and overall climate change, one of the major concerns is the loss of biodiversity, particularly on account of land use change. The expansion of land for soy mono cultures has led to the reduction of forest areas given that most soy in grown in monocultures. As it is more efficient producing biofuels in the South, considering a better climate and lower costs, biofuel production will be established in developing countries in the Southern hemisphere with countries with more land, longer growing seasons, lower costs of production. A major issue arising from soybean farming seems to be that the majority is produced through mono cropping which negatively effects soil in terms of nutrient depletion, biodiversity and exposes farmers in a dependence on soy bean/oil market prices. Despite the nutrient depletion of soy mono cropping, a switch to other crops is not directly feasible without significant investments which threatens particularly small and medium size farmers' existences. Thanks to the World Database on Protected Areas (WDPA)[214], areas representing a higher biodiversity and are vulnerable to climate change or human destruction can be registered and monitored. This register is used as official data to encourage (multi) national companies to avoid these areas for business purposes. For instance, only the Latin American country Argentina counts with over 300 terrestrial nationally designated protection areas among which are various affected by the agricultural expansion. The registered areas also include smaller locations, which are spatially visualized as point data, whereas the larger areas are visualized as shape data. As these point data cannot be properly visualized on a raster,

[214] WDPA, 2012.

only larger areas are subtracted from geographical potentials. Nonetheless, this subtracting of natural protection zones is not very efficient so that wildlife suffers in numerous areas. For instance, there are several "biodiversity hotspots" defined by the WWF,[215] which have been subtracted from the official list.[216]

In Argentina for instance out of the 300 areas, two concrete examples for such defined "no-go" areas are the *Yungas* and *Chaco* forest in Argentina.[217] The *Yungas* forest is located in the North western subtropical parts of South American country close to the Andean mountains. The area is situated on an altitude between 400 to 3,000 meters above sea levels and covers an area of approximately 5 millions of hectares (mha), which is one of the most diverse ecosystems in Argentina. The biological diversity is fragile and threatened by humans as higher levels of endemism comprise a significant number of unique plants and animals.[218] In contrast to the *Yungas* forest, the *Chaco* encompasses both dry and moist savannah ecosystems that extend over 70 million hectares in the Central and Northern part of the country. In *Chaco* there are high biologically diversified areas as well that are generally preferred areas for conservation since they are threatened by an increased exposure to being converted into agricultural land. In particular, governmental statistics show recently that in the provinces of *Chaco* and *Yungas*, approximately 250 kha of forests are being eliminated annually due to the purpose of soy. Seventy per cent of this deforestation happens in the Chaco areas. Further no-go areas in Argentina are for instance *Chaco Húmedo* and *Selva Paranaese*.[219] Finally these deforestations and destruction of natural ecosystems are driven by various factors comprising social, economic and political reasoning. Nonetheless, the expansion of soy cultivation is primarily caused by the continuously increasing demand of the livestock feed industry, which is dominant outside Latin America and requires the high protein SBM for hog, broiler and beef production in order to meet the rising demand of meet on international markets.[220]

---

[215] CI, 2012.
[216] Montenegro, 2011.
[217] Multinational Carbon Nanotubes for Biomedical Application, 2012 and CARBIO, 2012.
[218] WWF, 2003.
[219] Greenpeace, 2007 and 2011.
[220] GRAIN 2007d and FAO 2006.

3.8. Crop rotation and Genetically Modified Organism

Two aspects mainly determine such an expansion of farming land, crop rotation and Genetically Modified Organism (GMO). There is a natural crop rotation leading to higher yields so that farmers rotate between corn and soybeans on a five to ten years basis depending primarily on current market prices and their future expectations among other factors. In addition to these natural cycles, crop rotation mainly depends on expectations of price difference for soy and corn production. Crop rotation is necessary to balance the soil as different crops extract different nutrients from it. Concerns for the nutrient balance are often disregarded so that a deficiency of nitrogen, phosphor, potassium, and sulphur is caused through over-extraction by the same crop over years.[221] Especially SB need a lot of nutrients that most of the soil cannot provide alone anymore.[222] Thus, the use of fertilizers is of utmost important for farmers. These fertilizers are also a critical aspect within the soy cultivation as they add up to approximately 40 per cent of operating costs.[223]. When soils become over extracted and they loose all their nutrients, the land has toile fallow so that it can recuperate during which time farmers will look for more fertile arable land.[224]

In addition to crop rotation, GMO has an important stake within the expansion and coverage of new land. [225] In 2011, according to the USDA approximately 94 per cent of total soybean cultivation in the USA had been GMO,[226] and the California Department of Food and Agriculture even estimates that at least 70 percent of processed foods sold in U.S. supermarkets now contain genetically modified ingredients.[227] Whereas in Brazil, there are around 75 per cent of soybeans GMO, Argentina's GM share reaches up to 99 per cent. So that more than 90 per cent of Soybean cultivation globally is genetically modified in order to achieve more resistance towards droughts and diseases as well as higher yields leading to an increased efficiency.[228] The farmer's decision upon the use of GMO seeds depends mainly on the land and fertilizer use. When new land is converted

---

[221] De Souza, 2011.
[222] OECD, 2007.
[223] De Souza, 2011.
[224] Timilsina, 2010.
[225] World Bank, 2008a.
[226] USDA, 2012.
[227] Editora Gazeta, 2011.
[228] GMO-Compass, 2012a.

into arable areas and soybeans are cultivated for the first time, the farmer is better off, applying GMO as the seeds are more robust and resistant towards vermin, insects and plant diseases as well as needs less fertilizer. The amount of money he saves on fertilizer though will partially move to the crop science company that usually charges a commission on the final crop. Monsanto,[229] being a monopolist in Brazil for Soybeans directly provides seeds to farmers and charges the elevators or exporters for the purchased beans. This leads to a decrease in prices, the intermediaries are willing to pay and squeezes the farmer so that GMO will remain a highly discussed and controversial topic in upcoming years.[230] In order to accomplish the mandatory requirements of five per cent in 2010 and seven per cent, for instance, Argentina had to increase by nine per cent its farming area, referring to nine per cent out of the 16 mha,[231] which is estimated to increase up to fifteen per cent by 2023.[232] In this case, significant GMO improvements could change the rules of the game leading to more efficient land use and less overall farming area. These calculations only referred to nation-wide mandate increases so that external components including the rising demand in Europe and particularly China needs to be considered.

As in Figure 38, current biodiesel require a substantial share of total land demand. This share can be heavily decreased through the further progress in second and third generation biodiesel, moving from land masses to cultivation of algae and cellulose feedstock. Scientists do not consider GMO as a panacea to enhance yields (as indicated in Figure 39) and solve the land demand issue by its own, GMO is key in the advancement of R&D of current and future biofuels technologies. Nonetheless, the exponential increase in biofuel demand needs to be covered with an increased production.

[229] Monsanto, 2012.
[230] Simantob, 2011.
[231] GMO-Compass, 2012b.
[232] Greenpeace, 2007.

Figure 38: Biofuels demand and resulting land demand of the aforementioned IEA 2050 Scenario[233]

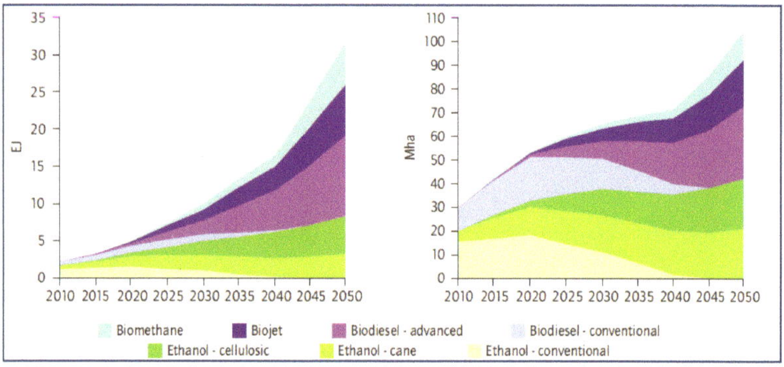

As the biofuel industry is heavily capital intensive, as outlined in the previous chapter, short term shifts are less likely to happen and political interventions need to be able to advance with more environmentally friendly and energy efficient technologies.

Figure 39: Land use efficiency development for biodiesel in 2010-2050[234]

Biodiesel	Yields in l/ha (2010) nominal	Liter biodiesel (lbd)	Average improvement per year (2010-2050)	Resulting yield in 2050 (lbd/ha)	Co-product 2010 vlaues (kg/l)
FAME Avge	2.000	1.800	1,0%	2.600	Glycerine (0.1)
RME	1.700	1.500	0,9%	2.100	RSM (0.6)
SME	700	600	1,0%	900	SBM (0.8)
PME	3.600	3.200	1,0%	4.800	Empty fruit bunches (0.25)

3.9. Direct (DLUC) and indirect land use change (ILUC)

In 2007, the World Resources Institute (WRI) published a report claiming carbon emission due to land use change can be a major factor in determining the overall climate impact of biofuels.[235] This is the displacement of other activities by biofuels production, causing deforestation and expropriation of farmers, communities and indigenous peoples. In general, the concrete definition of DLUC is straight forward: direct land use change is the conversion of land, which was not used for crop production before, into land used for a particular biofuel feedstock production. The emissions caused by the conversion process can be directly linked

[233] Based on IEA, 2011, 2010, and Figure 39, The gross land demand excludes land use reduction potential of co-products and is based on the assumption that 50 per cent of advanced biofuels are produced from wastes and residues.

[234] IEA, 2011, IEA, 2008, BRDI, 2008 and Accenture 2007. Biodiesel yields are presented as gross land use efficiency, without the land demand reduction potential through co-products. Density of Biodiesel is 0.88 and for simplification rounded to 0.9 t/m^3.

[235] WRI, 2009.

to the biofuel load and thus be allocated to the specific carbon balance of that biofuel.[236]

In addition to DLUC, ILUC is a market effect that materialises when biofuel feedstock are increasingly planted on areas already used for agricultural products. Ceteris paribus this causes a reduction of the area available for food and feed production and therefore leads to a reduction of food and feed supply on the world market. If the demand for food remains on the same level and does not decline, prices for food rise due to the reduced supply. These higher prices create an incentive to convert formerly unused areas for food production since the conversion of these areas becomes profitable at higher prices. This is the ILUC effect of the biofuel feedstock production. The ILUC effect of biofuels comes about only through the price mechanism of the global or regional food market.[237] Therefore ILUC in this context is always DLUC for food production incentivised by the cross-price effects of an increased production of biofuel feedstock which then translates into an additional demand for so far unused land areas.

Figure 40: Different types of land use change[238]

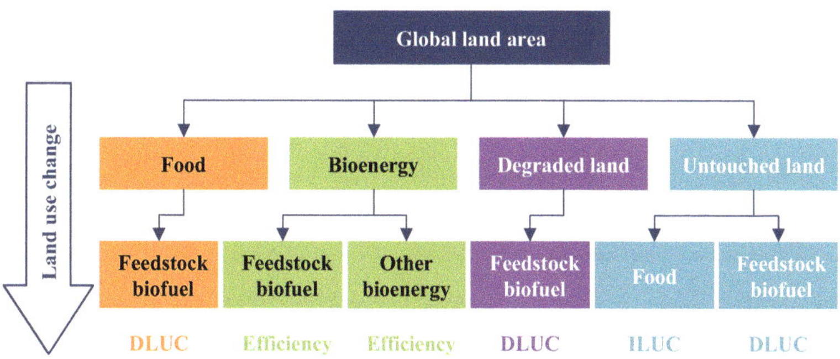

Figure 40 illustrates the concepts of different land use changes driven by an increased demand for biofuel feedstock. DLUC includes the conversion of food cropland, degraded land and untouched nature.[239] As described above, ILUC is a biofuel demand driven direct land use change of untouched nature for food production. We here consider also efficiency gains from biofuel production.

[236] Searchinger, 2008.
[237] Runge and Johnson, 2008.
[238] Delzeit, 2011.
[239] Delzeit, 2011.

Traditional bioenergy use, such as the use of firewood, can be replaced by biofuel feedstock or other bioenergy. This leads to efficiency increases as the advanced cultivation methods and technologies lead to higher energy yields per hectare. In particular this accounts for large areas in Africa, were crop based bioenergy production can replace tradition use of firewood. In order to assess the concept of ILUC one has to keep in mind, that land use can only be understood globally. The adjustment processes and the replacement of areas devoted to food production as well as the expansion of cultivated land areas is governed by complex global processes. Global demand and supply conditions as well as the regional support policies in the agricultural sector, local infrastructure and economic structure as well as the geophysical suitability of areas for agricultural production simultaneously determine land use decisions.[240] As a consequence, the ILUC of bioenergy production does not take place within the local community nor on a national level but come about throughout the globe. The casual linkage of a single driver/producer of biofuels to a particular indirect land use change is therefore impossible. Furthermore, the monitoring of these global processes requires highly spatially disaggregated remote sensing technologies which are still not available at global scale.[241]

The EU27 has just decided to delay any pronouncement on ILUC, yet it is clear that the issue is serious and urgently needs to be addressed. The provided data by the Commission do not include the net carbon emissions from land use change, assuming that no further emissions are generated by the direct or indirect land-use change.[242] Nevertheless, the land-use change and production technology influence GHG benefits for example in the case of the palm oil production. The palm tree plantations, particularly in Indonesia and Malaysia are heavily discussed and if a palm tree is grown on unproductive lands, GHG benefits can generally be very positive.[243] In addition, environmentalists have been urging the Commission to further introduce more sophisticated measurements of the GHG savings from biodiesel, arguing that the EU27 has to consider all the effects of ILUC.[244] These could be in the form of displacing food crops by feedstock cultivation or expand-

[240] Montenegro, 2011.
[241] Delzeit, 2011.
[242] Delzeit, 2011.
[243] IEA, 2011.
[244] European Voice, 2011.

ing farming areas to previously uncultivated land that might, for instance, host environmentally valuable woodland or one of the aforementioned no-go areas.[245] In spite of the more than 150 international scientists warning that the indirect land impact of biofuels was significant and that current scientific understanding justified immediate action by EU policymakers, the European Biodiesel Board (EBB) published a position paper emphasising as well on the need of competitive feedstock from South America and south East Asia.[246] The Institute for European Environmental Policy (IEEP) notes that just for EU biofuels under the Renewable Energy Directive target by 2020: "Using currently available data, this additional demand for these fuels is anticipated to lead to between 4.1 and 6.9 million ha of ILUC, that is, an area equivalent to just larger than Belgium to just under that of the Republic of Ireland."[247] The Commission takes into consideration only one aspect of sustainability, namely the environmental aspect. Even within this aspect it omits including the effect of changes in land-use. However the sustainability cannot be interpreted from only one aspect, it is mostly a combination of environmental, social and economical issues. In the upcoming years, the EC aims to include all of these aspects in the sustainability criteria.

Even external consulting companies like McKinsey refer to a recently published study about resources,[248] highlighting the comprehensive understanding of integrated resource benefits as well as the correlated capital and costs. Thus, in addition to water and energy use, direct and indirect land use is one if the essential levers to drive carbon abatement efficiently.[249] Thus, competition for land ends up in another key point for the sustainability of biodiesel. In general, land should always principally serve for food production. While the efficiency of food production should be increased, feedstock for energy generation should only be planted on surplus land,[250] avoiding the threat of deforestation or loss of biodiversity. The logging industry expands and unsystematic cutting of primary forest is still a common practice in developing countries for the extension of farm land. The prevention of an uncontrolled extension of farm land for crop plantations is a

245 European Voice, 2011
246 EU, 2010.
247 IEEP, 2010.
248 McKinsey Global Institute, 2011.
249 Dobbs, 2011.
250 Smeets, 2004.

major concern in the development of a sustainable biodiesel trade, particularly given that current technology still depends on first generation biodiesel.

Finally, ILUC is not a unique phenomenon within the area of biofuels or in relationship with specific geographic regions. Any form of additional demand on the global agricultural system has the potential to create ILUC. Additionally, biofuels account for a very small proportion of global agricultural production; around two per cent, or 36 mha,[251] from a total cropland area of around 1,527 mha.[252] So, non-agricultural land use changes may also result in ILUC impacts, such as urban expansion or infrastructure development. Therefore the most effective means of mitigating ILUC is likely to be effective, global regulation on land use. This is beyond the remit of the private sector but nonetheless should remain an important consideration for policy makers and political frameworks as it will be outlined in the next chapter.

[251] UNEP, 2009.
[252] FAO, 2012.

4. Political frameworks and incentives for the biofuel industry

During the past decade, biofuels appeared to offer a cleaner and overall greener alternative to fossil fuels, leading to economical and political support. There are a numerous reasons to promote biofuel trade globally. On a local level there is the generation of rural development, on a national it can be a source of increased energy security and enhanced energy diversification and on a global level it has the potential to reduce GHG emissions[253]. Furthermore, an increased production or import of biofuels provides energy security versus fluctuating oil prices and oil cartel structures. Additionally, biofuels bear an alternative energy source as domestically produced, offering the opportunity of reducing dependence on imports of energy sources.[254] Moreover, biofuels also create possibilities for diversification of the agricultural industry and the opportunity of a value-added product. Establishing bio energy programs in rural areas have the potential to generate development for small-scale farmers by the government or processing companies. This could further generate investments and broaden the scope of business activities. However, as in the previous chapter outlined, potential impacts question political interventions and challenge the global community to develop new frameworks for successfully alleviate upcoming climate change consequences. In addition to concerns about GHG emissions, other impact areas in relationship with biofuels arise.[255] Firstly, food prices and food security, particularly in poor households within food-importing developing countries are affected by biofuels.[256] Secondly, economic distortions and unintended consequences of political interventions including OECD subsidies lead to shifts in micro and macroeconomic environments. Finally, the biofuel industry directly and indirectly causes environmental impacts affecting biodiversity, land-use changes as well as air and water quality.[257] Hence, climate change and the above mentioned consequences can be considered as market failures and require public sector intervention. These political instruments can vary in form of scope and scale but can be

[253] IEA, 2007.
[254] Bauen, 2009.
[255] Taheripour, 2010.
[256] OECD, 2008.
[257] Pelkmans, 2008.

commonly summarized within five major policies comprising tax reductions, trade tariffs, mandates, direct and indirect subsidies.[258]

Firstly, governments can stimulate an industry sector with tax reductions in order to regulate prices creating economic incentives. Argentina for instance introduced this kind of fiscal strategy on the biodiesel export taxes in order to encourage the biodiesel industry to invest in refinery capacities within the country as described with more detail in the next chapter (4.1).[259] Secondly, positive or negative import tariffs as for example a volume based tax credit implemented in the US in 2004 lead to a sharp increase of biodiesel into the country.[260] The so called splash and dash effect created by this fiscal policy will be described in Chapter 4.2. Another tariff system is implemented by Morocco on commodities whereas products from the US enjoy from lower import tariffs so that the comparatively more expensive soybeans (SB) or soybean meal (SBM) can be imported on a more competitive price than the Argentinian equivalent, which would be cheaper in the Mediterranean area.[261] Complementary to tariffs, (quality) standards can be introduced as for instance the sustainability standards requested by the Fuel Quality Directive (FQD) in line with the Renewable Energy Directive (RED). In Chapter 4.3, more detail on the European requirements will be elaborated so that a clear understanding about this kind of political intervention can be gained. The third political measure is the setting of mandates as already highlighted in Chapter 2. The EU27 clearly uses them in order to encourage member states to comply with the 20:20:20 Energy Strategy.[262] These mandates are also sometimes described as "renewable fuels standards" foster the biofuel industry in cases of bioethanol and biodiesel to establish strategic investment by guaranteeing a certain domestic demand. These mandates are usually increased on a continuous basis in order to cope with production capacities and aim at long term industry stability avoiding shortages as well as higher price volatility. At fourth place are direct biofuel production subsidies, allowing the entry of new players within the industry as parts of the costs can be covered through them. Unfortunately, subsidies are also likely to increase input costs and in case of biofuels, rising

[258] Harvard University, 2007 and Lawrence, 2010.
[259] Gallo, 2011.
[260] UFOP, 2011.
[261] Guerrero, 2012.
[262] EREC, 2008.

feedstock prices. In order to compete with cheaper areas particularly in the southern hemisphere and to pursue other political incentives like the reduction on energy dependence from Russia and the Middle East or the support for the domestic agriculture industry, the EU27 provides massive subsidies for Rapeseed production and biodiesel refineries. Especially due to the rising feedstock prices on global commodity markets pushed by positive macroeconomic outlooks and financial activities, biodiesel producer would have produced on a loss without the economic incentives given by the public sector.

Figure 41: Comparison of international market prices of diesel and vegetable oils[263]

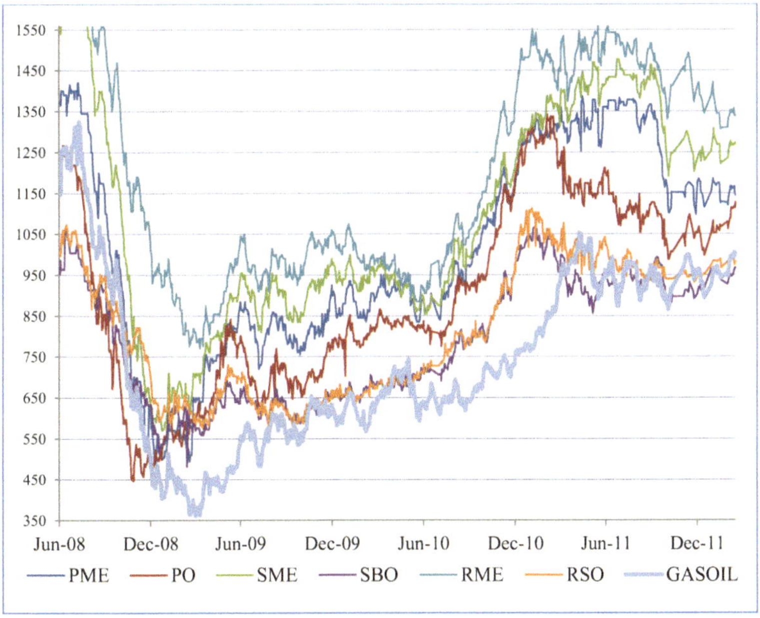

As Figure 41shows, the feedstock prices, for instance vegetable oils (PO/RSO) in Rotterdam are already higher than the equivalent diesel price (gasoil). At current vegetable oil prices, it would be also cheaper to import FAME directly from Argentina or Malaysia in form of SME and PME accordingly than importing the oil and refining it locally.[264] As biodiesel economies in Europe have been more unfavourable and the operating costs including feedstock, overheads and produc-

[263] Own demonstration based on Noble Group Research 2012b. The gasoline price is represented with the Intercontinental Exchange gasoline future price whereas the vegetable oils and methyl esters are given as the Rotterdam Free on Board (FOB) price.
[264] World Bank, 2010.

tion have been higher than the final product itself, European governments subsidise the industry so that domestic production and trust into the industry sector can be maintained as well as further investments is encouraged. In addition to production subsidies, further support can granted in in form of distribution, storage and biofuels transport subsidies.

All these economic incentives have been on the biofuels supply side and can be complemented by the introduction of a demand-side subsidy for instance for car manufacturers encouraging them to build more fuel-flexi vehicles as well as a tax reduction targeting customers who purchase more environmentally friendly cars. Creating incentives that intent to generate a higher demand can be considered as an indirect subsidy. Another industry support exists in Argentina where the government provides export licences for certain commodities including corn and wheat.[265] In order to control the exported volume and therefore protect the domestic market including the livestock industry, guaranteeing them a sufficient grains supply, the government restrictively conveys these licences. In this way the demand side is therefore kept within a certain limit leading to lower domestic prices as a macroeconomic consequence which indirectly supports the local feedstock industry and provides an income stream for the government.[266]

Other policies can be implemented in form of financial structures enabling better financing and investment access regarding grants and loans as well as the introduction of minimum prices securing a certain margin for producers, among others.[267] Finally, there is not a single measure to fit all circumstances and being the solution for a sustainable development. Most likely it needs a combination of fiscal and monetary policies in order to satisfy all stakeholders involved. Various policy instruments are used within the EU to promote biofuel use, namely fuel tax exemptions for biofuel producers, blending or use targets, trade measures in terms of import tariffs and sustainability standards in order to stimulate sustainable development as well as efficiency in the supply chain.[268] As mentioned in Chapter 2, the RED encouraged each MS to implement national blending targets of at least

[265] O'Farrell, 2012.
[266] Gallo, 2011.
[267] USDA FAS, 2009d.
[268] 2BSvs, 2012.

2 per cent by the end of 2005, rising to 5.75 per cent by end-2010 and reaching the individual objectives as in Figure 5 in Chapter 2.1 by 2020.[269]

4.1. Argentinian's export tax driven biodiesel industry

In Brazil a considerable amount of produced SBO supply local biodiesel producer that focus on the domestic market providing biodiesel to crude oil refiners and distribution companies. In contrast, the Argentina established favourable conditions for SME export[270] with the introduction of an incentive mechanism, which included tax exemptions for biofuel producers and legal blending requirements from the beginning of 2010. The initial mandate of 5 per cent for diesel[271] has been increased to 7 per cent and is likely to be elevated up to 10 during the following five years.[272] This export-driven strategy-tax incentive refers to biodiesel production, as an initially standardized product in form of agriculture commodity benefits from tax reductions while first converted within a value added process and then exported.

In general, taxes on fuels are considered as important income stream government among indirect fiscal policy instruments, especially since fuel consumption is comparatively inelastic.[273] While agreeing upon a fuel tax reduction as economic incentive to promote biofuels, the government will reduce its revenue streams. Nevertheless, a small reduction might already trigger the industry to invest in infrastructure and further expand production capacity. Instead of the common value-added tax (VAT) of 21 per cent on revenues,[274] the export tax varies from product to product. The export tax for agricultural products is on average 20 per cent of the Free on board (FOB) value, whereas for corn 23 and wheat 20 per cent.[275] This relatively high tax for soybeans of 23.5 per cent encourages local crusher to process the beans and export only the products. Thus, more SBM and grains in general remain within the country and represent a stable supply for the feedstock industry including broiler, hog and cattle farms. Moreo-

[269] EC, 2009.
[270] USDA FAS, 2009a.
[271] World Bank, 2010.
[272] Gallo, 2011.
[273] Bacon, 2001.
[274] Gallo, 2011.
[275] Moltoni, 2006.

ver, the SBO and SFO export tax ensure that local food supply is given and food prices are kept on a low level.[276]

Figure 42: Example for export taxes for agricultural products[277]

Commodity	SB / SFS	SBO / SFO	Grains	Biodiesel
Tax (in % of Export value)	23.5%	23.5%	20.0%	5.0%

The difference in exports tax of vegetable oil and biodiesel are 18.5 per cent (as in Figure 42, 23.5 and 5.0 per cent accordingly). This discount of biodiesel is a strong incentive driving the vegetable oil producers and new investors to expand vertically their supply chain capabilities up to transesterification facilities. That is why over the last decades, large investments have been made. Most of these new plants have been constructed along the Rosario belt directly connected to the River Plate, which altogether represents a convenient location in terms of logistics so that soybean feedstock can be easily sourced internally and exports can directly be loaded from the crush or refinery plant via the integrated port facilities on the seaside. Thus, Argentina could develop a competitive advantage and increase the approximately 400 kmt of SME exported in 2008 up to over 2 mmt in 2011.

4.2. Splash and Dash - a US case study on import and tariffs

In 2009, the EU put anti-dumping and countervailing duties upon US biodiesel imports in order to offset the so-called 'splash-and-dash' practice or B99 effect. The US Congress had established a volumetric excise tax credit (VETC) for biodiesel, which was blended with fossil fuel so that imported biodiesel to the US from a third country (e.g. RME from Canada or SME from Argentina), could be re-exported while obtaining the tax credit. This practice was commonly known as 'splash-and-dash'. These re-exports exclusively went to the EU27 zone where the biodiesel would receive a second financial incentive through different MS's support schemes (Figure 44).

[276] O'Farell, 2012.
[277] Own demonstration based on Agroparlamento, 2012 and Gallo, 2012.

Figure 43: US foreign trade balance for biodiesel

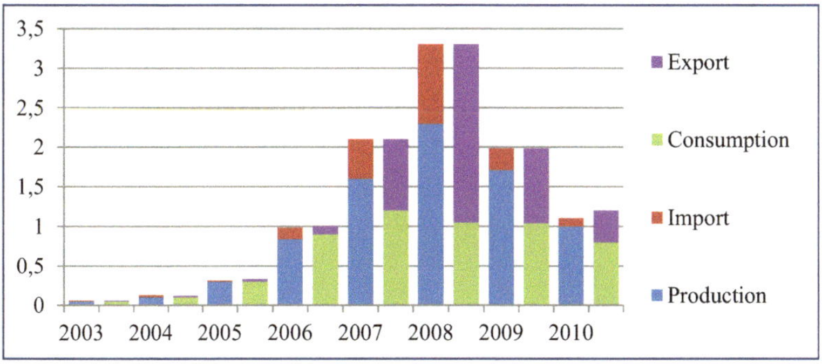

The term 'B99 effect' came from the fact that the definition of 'blending' made it possible to receive the credit by adding a mere 0.1 per cent of diesel resulting in trade of B99.9 biodiesel. As in Figure 43, the US imported biodiesel despite their production surplus so that this rising import demand did not correlate with the consumption or storage but was exclusively directed to be re-exported.[278]

Figure 44: Global biodiesel trade streams in 2008 (kmt)[279]

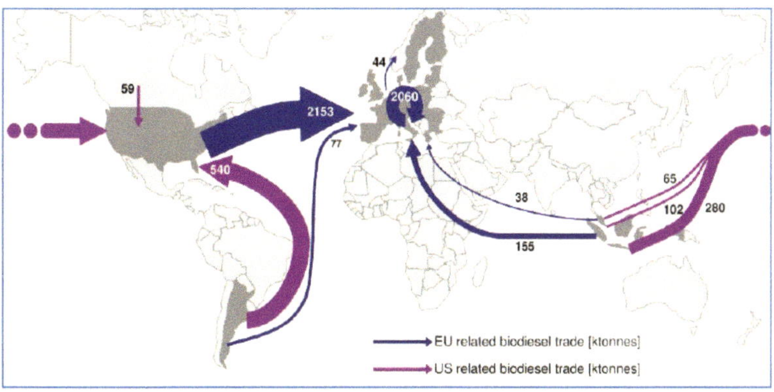

The volumes and trade flows as well as their changes after the European intervention are demonstrated in Figure 44 and Figure 45 respectively. Argentinean exports to the US increased sharply in 2008 and reached a climax of the B99 effect, but dropped again in 2009 when the EU27 introduced these counter duties. This pattern is also true for Indonesia and Malaysia in terms of PME. As the

[278] Vasquez, 2012.
[279] UFOP, 2011, based on Lamers, 2011.

VETC was extended until the end of 2009, the EC imposed antidumping and countervailing duties on US imports, effective as of July 2009 under the Regulations 598/09 and 599/09,[280] which reduced direct US imports significantly (Figure 45). A formal investigation of this issue was launched by the EC in August 2010 following a complaint from the EBB,[281] which led to an extension of US focused anti-dumping and countervailing duties and the creation of such measures against imports from Canada.

Figure 45: Global biodiesel trade streams in 2011 (kmt)[282]

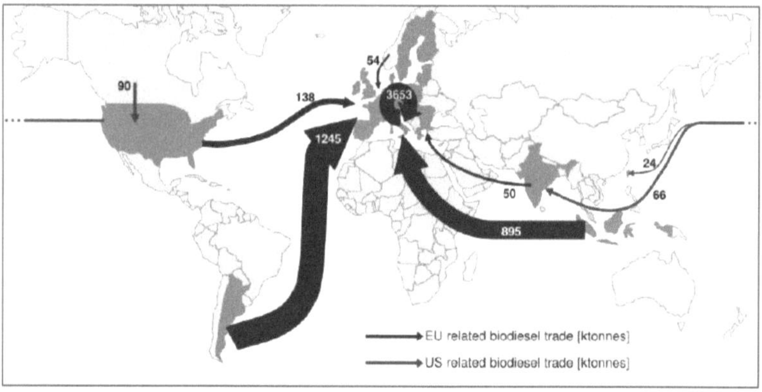

4.3. Sustainability certificates and voluntary schemes – 2BSvs case study

The EU Renewable Energy Directive (RED) demands that only certified sustainable biofuels are eligible for tax benefits or may be considered for the 2020 objectives. That is why shortly after the RED introduction in 2009, several independent auditors and certification bodies have been developing voluntary schemes for the actors along the biofuel supply chain. There are currently seven certificates and voluntary schemes that have been officially recognized by the EC. There are the International Sustainability and Carbon Certification (ISCC)[283] that was financed by the German government, the Biomass Biofuels Sustainability

[280] UFOP, 2011 and EC, 2009.
[281] EBB, 2011.
[282] UFOP, 2011, based on Lamers, 2011.
[283] ISCC, 2012.

voluntary scheme (2BSvs),[284] which is a French industry scheme and the Roundtable on Sustainable Biofuels (RSB)[285] which was initiated from Energy Center of École polytechnique fédérale de Lausanne. These three cover all types of biofuels. Additionally, the Round Table on Responsible Soy Association (RTRS)[286] is literally a roundtable initiative for soy-based biofuels with a focus on South America. The other three certificates focus exclusively on ethanol production and are therefore related to sugar cane and corn feedstock. Whereas the Bonsucro[287] is a roundtable initiative for sugarcane based biofuels with major focus on Brazil, the Abengoa Renewable Energy directive Bioernergy Sustainability Assurance (RSBA)[288] is an industry scheme covering the supply chain of the American-Spanish bioethanol producer Abengoa. Finally, Greenergy[289] is also an industry scheme designed for the firm Greenergy covering their sugar cane and ethanol production in Brazil. Thus, biodiesel and bioethanol made from soybean, rapeseed and sunflower oil as well as corn and sugar cane are already covered with certificates, while a certificate for palm oil is not yet approved by the European Commission. Nonetheless, the Round Table on Sustainable Palm Oil (RSPO) is in its final stages of being recognized by the public authority in Brussels.[290]

The Voluntary Scheme 2BSvs was developed through a consortium of major agricultural and industrial actors of the French biofuel industry which was supported by *Bureau Veritas*, an international leading assessment and certification auditor.[291] The scheme's objective is to demonstrate compliance, via an independent audit made by a certification body, with the RED sustainability criteria. After having been successfully assessed, biomass used as feedstock or produced biofuels become certified and therefore relevant to the Energy Strategy 20:20:20 objectives (as illustrated in Chapter 2.2). This concrete sustainability scheme is based on a voluntary audit methodology aiming at different stakeholders along the biofuel supply chain. First, it addresses farmers, crop producers and agencies

[284] 2BSvs, 2012.
[285] RSB, 2012.
[286] Round Table on Responsible Soy, 2012.
[287] Bonsucro, 2012.
[288] Abengoa, 2012.
[289] Greenergy, 2012.
[290] RSPO, 2012.
[291] Bureau Veritas, 2012.

responsible for the collection of biomass. These agencies are either elevators who both, physically own and store the raw material, or who just sell their storage services to producers and trading companies.[292] Secondly, agricultural cooperatives and agribusiness companies which are involved in the vegetable oil production including large corporation that are vertically integrated like for instance Bunge that owns soybean plantations, elevators, crushing and refinery plants as well as port facilities.[293] Furthermore, it certifies processors and traders who purchase and resell either the feedstock or the processed products and therefore physically own the biomass along the value chain.[294] The voluntary scheme covers all sustainability criteria which are required by the RED along the supply chain independently whether the feedstock is cultivated within the EU27 or imported from outside the Schengen area.[295] In order to be 2SBvs certified, several entities needs to comply with the sustainability criteria as already presented with the RED in Chapter 2.2. Foremost is the reduction of GHG emissions compared to conventional fuel.[296] Furthermore, biomass shall not come from lands with high biodiversity values including primary forests, protected areas and ecosystems, high biologically diverse grassland or land with high carbon stock such as forested land, wetlands and peatland.[297] All MS shall define areas to be excluded, or submitted to conditions, on behalf of biodiversity.[298] In order to track the quantity of sustainable biomass, a MBS should be implemented for each logistical site as well as a control system for each economic operator, that is, each legal entity that owns the raw material or finished products along the supply chain. These economic operators shall be able to demonstrate that sustainability criteria are fulfilled under its verification scope via the implemented monitoring and control system. Each entity should be audited by an officially approved and independent verification body. The control and monitoring system comprises on the one side procedures, instructions, training, information for the operator's location and staff as well as on the other side proofs demonstrating that all scheme

[292] Simantob, 2011.
[293] Montenegro, 2011.
[294] 2BSvs, 2012.
[295] Bureau Veritas, 2012.
[296] Montenegro, 2011.
[297] An area in which peat occurs naturally; peat is a soft black or brown substance formed from decaying plants just under the surface of the ground, especially in cool wet areas. It is burned as a fuel or used to improve garden soil; Oxford Learners Dictionary, 2012.
[298] 2BSvs, 2012.

requirements are fully met and controlled. Moreover, the MBS is supposed to be implemented for each location, including sub contractors and should ensure that the quantity of sustainable material sold does not exceed the amount of material purchased at any given moment. Hence, traceability of sustainable material along the entire supply chain is given. At least once a month, the mass balance must be verified to ensure positive balances.

Case Study: Project Outline of the 2BSvs Implementation[299]

Figure 46 illustrates a time line of the Noble Group case study below referring to the project overview for the 2BSvs implementation in 2011/2012. The outline demonstrates how a theoretical audit approach can be transferred into practice whereas in the next chapter, challenges and problems of such an implementation will be highlighted. Noble Group is a Supply Chain Manager and reaches a global scale with its operations encompassing a total of 150 offices in 38 countries. As a Fortune 500 company based in Hong Kong (China) and listed in Singapore, the firm operates regional headquarters in Geneva (Switzerland), London (UK) Stamford (USA), Sao Paulo (Brazil) and Buenos Aires (Argentina). Currently the company employs around 12,000 employees and generates an annual turnover of US$60 billion with a net profit of approximately US$500 million whereas the agriculture products segments represent approximately one third of their business activity besides other sectors such as energy and logistics.[300] Noble Argentina, a Noble Group subsidiary and vertically integrated commodity trader in Argentina operates a crushing and SBO extraction plant as well as a directly connected port facility in order to export SBO and SBM products overseas.

[299] Based on a real case scenario of Noble Argentina, 2011.
[300] See Appendix for Noble Company presentation

Figure 46: Timeline of the sustainability certification project in Argentina[301]

Time line		2011			2012	
Activity	**Oct**	**Nov**	**Dec**	**Jan**	**Feb**	**Mar**
NASA Timbues						
Establishment of documens						
Reception & MBS						
Verification						
Sales - Sustainability certificate						
operational Changes (IT, proceses, etc.)						
Reception of Mass (Sustainable SB)						
Supervision of areas and producers						
Monthly and annual reports						
Selection of preferred suppliers - who do we include?						
Producres (not to be certified)						
Sign the declaration of conformity						
Establishment of mapping (third party: Yungas)						
Elevators / Brokers (certified as First gathering point)						
Elaborate commercial offer as an incentive						
Same producers as for NASA Timbues						
Training of people						
Determine one person in charge / responsabilities						
Sales - Sustainability certificate						
Audit of the plant						
Employ exernal consultant						

Time line		2011			2012	
Activity	**Oct**	**Nov**	**Dec**	**Jan**	**Feb**	**Mar**
NASA Buenos Aires / Rosario						
Establish contact with ke suppliers						
Approve mapping request						
Negotiation with suppliers						
NRSA Geneva						
Establishment of documens						
MBS						
Sales - Sustainability certificate						
Operational changes						
MBS						
Sales - Sustainability certificate						
Training of people						
Determine one person in charge / responsabilities						
Audit of the Broker						

As every part of the supply chain (that physically owns the soybeans) has to be certified so that both, the crushing plant in Timbues as well as the office in Buenos Aires of Noble Resources SA had to be certified. Argentina started working with the producers that cultivate soybeans and directly supply them to

[301] Own demonstration based on project lead for Noble Group's sustainability certification

their plant. Instead of an individual certification and audit, they only needed provide a signed conformity declaration,[302] stating that their areas are considered as "go" areas by the Argentinean Chamber of Biofuels.[303] These "go" criteria are in line with the RED.[304] In addition to the declarations, images or maps that confirm their areas being located in these approved areas had to be elaborated to the certification body. After having elaborated both, Noble Argentina could start with the auditing process itself.

Firstly, Noble Argentina's operation plant in Timbues needed to be certified and documents (in Spanish) describing all the procedures which had to be established. These procedures include the reception of the sustainable raw material as well as their tracking via a Mass Balance System (MBS) with which it is possible to determine the exact amount of stored sustainable soybeans at any time. Moreover, the verification process that involved the declaration and mapping of suppliers has to be developed so that every quarter, an update of "go" and "no go" areas according to CARBIO could be made. These updates also track whether new or old producer change or expand their areas so that a constant control of origination locations can be guaranteed. Finally, the sales process including the establishment of a sustainability certificate has been documented. This certificate states the certification number, date, volume, product, GHG emission saving and vessel name of according shipment. Based on the documents, changes within the Information technology systems could be carried out so that the reception of sustainable soybeans is automatically tracked and is directly linked to the MBS, which cannot reach a negative balance. The supervision of areas and producers in terms of "go" and "no go" areas is tracked on a regular basis so that the sustainability of the raw material can be guaranteed at any given time. Thanks to the MBS and the regular updates, monthly and annual reports can be easily published so that the reporting complies with the certification requirements. As the amount of suppliers is quite elevated, the selection of a first sample of suppliers, with which Noble Argentina wanted to start implementing the voluntary scheme, was based on a commercial decision. Generally, the certifica-

[302] 2BSvs, 2011.
[303] Camara Argentina de Biocombustibles (CARBIO), 2012.
[304] European Parliament and Council, 2009.

tion scheme differentiates producers from elevators in that the first need to sign only the declaration of conformity while the latter has to be separately certified as a first gathering point. In Argentina, like probably in most of the countries worldwide such a certification or signed declaration can only be set in place via a financial incentive. That is why Noble's need for signed Declarations of Conformity providing and GPS data of the field locations was only possible through negotiations. Thus, origination directors and the country director elaborated a commercial offer as incentives for the preferred suppliers. Based on the reception of the Declaration of Conformity including the GPS data a mapping confirming the producers' SB origin was feasible. Additionally to the producing supplier, commercial offers for elevators and brokers have to be elaborated in order to leverage the certification and reach higher volumes. In order to simplify the audit, Noble investigated which or the first gathering points were already implementing 2SBvs or any equivalent scheme or which one at least demonstrated such an interest in the market before approaching the ones with which they had strong relationships. Finally, the training of people regarding the scheme implementation has to be organized and a training programme was elaborated. In addition to the trainer and person responsible for the education, a person had to be declared as in charge of the certificate. This person is also the contact point of the certification body.

Secondly, the Noble Argentina offices in Buenos Aires and Rosario a well as Noble Resources in Geneva had to be certified since they own the raw material during the conventional trading cycle. Noble Geneva operates as a broker and commodity hub because the soybeans are generally purchases directly from its subsidies in Argentina and Brazil and then resold, for instance to other market participants in Europe. In terms of costs, the mapping of the producers arose to approximately US$40-60,000, depending on the exact GPS data from producers, that is, with each GPS data Noble got directly from the producers or elevators via the Declaration of Conformity, ARG$500 had been saved. In addition, there were the cost for the certification of the plant (NASA) of the Broker (NRSA) and additional costs for economic incentives to third parties including the Producers,

Brokers and Elevators as well as additional upcoming costs for every auditory update in the next years.

4.4. Criticism on sustainability schemes

Especially the rapid scaling up of biofuel production in many countries calls for immediate action and requires a uniform certification scheme. Collaborative certification schemes set minimum standards for the oilseed cultivation and harvesting practices as well as the feedstock and biofuel production process.[305] Since biofuels trade has been developing and has become more complex reaching higher volumes on a global scale, more advanced and sophisticated certification schemes need to be introduced.[306] Among the seven afore-mentioned schemes in Chapter 4.3, there have been other certificates that have been proven as efficient and served as role models. The Rainforest Alliance's Standard for sustainable Agriculture in Latin America that already spans a network of organic crop cultivation in more than 100 countries intents to improve the social and environmental conditions in tropical regions. Founded by the Sustainable Agriculture Network (SAN)[307] the certificate aims to include all stakeholders within the agriculture industry to take on more responsibility and thus, close the North-South gap through joint initiatives along the supply chain. Within ten key areas approved by the SAN, the conservation of ecosystems, in particular wildlife and water as well as fair treatment, waste and environmental management had provided a base for the Biomass and biofuels standards we have nowadays.[308] The Forest Stewardship Council (FSC), headquartered in Bonn, Germany, is an international forest certification system network of non-profit organizations (NPO), businesses and governments aiming at the promotion of responsible forest management worldwide in particular emphasizing on minority and indigenous groups as well as forestall benefits jeopardized by impacts of logging and poor forest management.[309] The FSC created a basis for the Biofuels Sustainability

305 Montenegro, 2011.
306 Vasquez, 2012.
307 Sanstandards, 2012.
308 Rain Forest Alliance, 2012.
309 FSC, 2012.

certificates as most of the forest and highly biologically diversified areas in Latin America are threatened by agriculture activities.

However, current discussions about standardization bear doubts as for instance whether biofuels should held higher standards than agricultural products or comparatively low one as for petroleum based conventional fuels, particularly referring to countries where the same crop is used for biofuels feedstock and for the food industry, as for example in Argentina and Brazil where Soybeans are used as biodiesel feedstock and for the food industry (as in Figure 13). Lobbyists and protectionist of the agriculture industry argue whether sustainability standards and certification schemes support or slow down the pace of market development, especially in developing countries where lack of infrastructure and transparency among public authorities form part of daily challenges. Noble Group has had substantial problems to convince farmers about the advantages of such a certification scheme. Farmers and elevators had not been ready to sign any declaration as they are generally sceptical towards information provision. This mistrust is mainly triggered by the Argentinian authorities. Due to a lack of transparency, farmers had already lost their trust before public and even some independent authorities.[310]

Furthermore, standards should not only focus on technical features like the GHG reduction or the origination location, but rather have an overall approach as in Figure 47 including social and environmental aspects.[311]. Currently, the focus lies primarily on big scale production of biodiesel derived from SBO. However, a more balanced development regarding production scale would be favourable, that is, market access to small and medium size farmers, and a more diversified feedstock production. Nonetheless, the sustainability criteria are for most of the smaller farmers an additional obstacle leading to comparatively massive administrative and bureaucratic difficulties with which they have to handle. As already mentioned in the previous chapter, even for Noble Argentina, which is one of the top ten players in the domestic soybean business, the certification process was a bureaucratic and administrative challenge.

---

[310] Passerini, 2011.
[311] Gallo, 2011.

Figure 47: Sustainability aspects framework affected by biofuels[312]

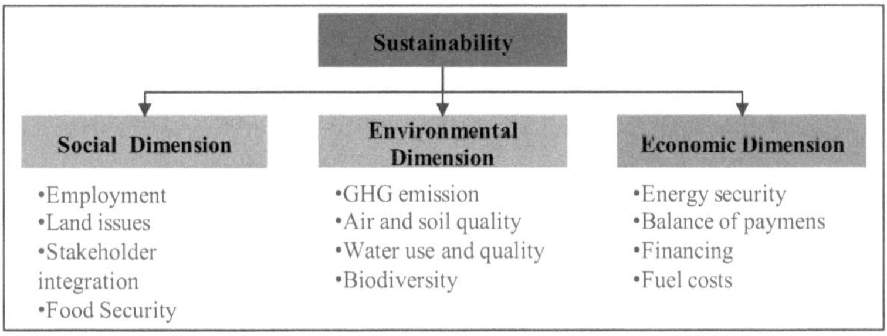

According to critics, the EC created this sustainability scheme not only because of sustainability consciousness in the fight of climate change but also as a market protectionism mechanism for its agriculture and biofuel industry, which have been suffering under low cost imports from particularly Argentina and Indonesia in terms of SBO and SME as well as PO and PME accordingly.[313] Thus, the introduced sustainability criteria in combination with European biofuel quality standards enable to protect local rapeseed farmer and biodiesel producer operating with rapeseed that has been grown in Europe and prevent to a certain degree cheap raw materials of being imported.[314] In general, the introduction of such certificates and standards has to be assessed whether it supports a sustainable development, fostering companies to truly invest in order to meet the criteria and therefore lead sustainably to an overall improvement in terms of environmental and social objectives or whether the certificates hinder a possible more accelerated climate change counter scenario. In addition, the ambitious GHG reduction targets and biofuel mandates could even be jeopardized by the suspiciousness created through overregulation.[315] Currently, there are few sustainable oilseeds available from outside Europe, yet. On the contrary, the entire 2010 German domestic rapeseed production had already obtained the certification. As a result they were destined into biofuels production, while rapeseed oil for food use was imported.[316] The regulation does not require a sustainability standard or certificate for the food sector yet. However, some NGOs push for broadening the scope to

[312] Own demonstration based on MSc programme at HEC Paris and IEA, 2011.
[313] Simantob, 2011.
[314] Oxfam International, 2011.
[315] Passerini, 2011.
[316] Kabath, 2012 and DBFZ, 2011.

food and the feed industry, too.[317] This sustainability standard extension would be reasonable as soybean cultivation is mainly driven by the protein demand for SBM and disregards the comparatively cheap SBO.[318] In addition, some food companies are working on an officially recognized sustainable palm oil certification according to Roundtable on Sustainable Palm Oil (RSPO).[319] Besides, *Bunge* and *Louis Dreyfus Commodities* among other supply chain managing companies spread actually rumours to extend the Sustainability Origination of Soybeans to produce Soybean Oil to Soybean Meal that is mainly used to feed livestock.[320] After having reached a mature, such a certification would not require much more hassle for companies that already have their origination certified stage as the soybean crushing plants produce SBO and SBM at the same time. Furthermore it could be an additional incentive triggering the industry to invest in certification processes.[321]

In Europe, the vegetable oils and protein meal industry spans a network with more than 150 facilities providing to approximately 20,000 people employment. More than 30 mmt of basic products are yearly produced which are directly supplied to the animal feed-, pharmaceutical-, chemical-, food- and cosmetic industry as well as to crude oil refineries as seen in Chapter 3.2.[322] Represented by FEDIOL (European Union Oil and Protein meal Industry), this industry achieves a considerable weight in their lobby activities. In addition, COCERAL (Comité du Commerce des céréales, aliments du bétail, oléagineux, huile d'olive, huiles et graisses et agrofournitures)[323] is the representative organ for the European cereals, rice, oilseeds, olive oil, oils and fats as well as agricultural collectors, distributors, exporters, importers and agriculture-bulk stores of the according commodities from the EU-27 Member states.[324] FEDIOL and COCERAL criticize the current state of regulation in relationship with the sustainability certification schemes in a

317 USDA, 2011g.
318 Guerrero, 2012.
319 RSPO, 2012.
320 Vasquez, 2012.
321 Kabath, 2012 and Zylla, 2012.
322 FEDIOL, 2012.
323 COCERAL, 2012.
324 COCERAL, 2012.

twofold way. First, the Schemes validity and recognition is not uniform and second the MBS time period is too short.[325]

Even though the legislative framework recognizes the equivalence of different schemes, there is no consensus between the individual scheme owners in term of mutual recognition that generates market confusion and potential for trade distortions in production and physical flows. The validity lack creates a risk along the supply chain so that the existing flexibility of cooperation between suppliers and operators is jeopardized.[326] The standardized commodity market will experience a reduction in operator's freedom to decide upon the most adequate scheme and therefore leads to hindering the availability of agricultural products and even up to supply shortages.[327] The variations in time periods between the certificates and voluntary schemes create an additional source for supply chain risk. Discrepancies of MBS validity periods between three and twelve months should be removed. FEDIOL and COCERAL address the EC to uniform the MBS periods in order to create a solid basis of twelve months for a functioning supply chain. In particular Argentinian farmers lack of trust in legal entities and are discouraged by audits so that hey tend to perceive a three-month period as a prescriptive constraint.[328] Only a top down approach from the EC towards its MS, related scheme owners and certification bodies encouraging them to embrace such scheme equivalence as well as uniformity could convey a sustainable development. Thus an acceptance by the EC should automatically bear the recognition by all 27 EU MS governments.[329]

Nonetheless, the current standards are not at all a insurmountable obstacle for actors along the supply chain and on a short time basis if a producer or trading house is not be able to comply with the necessary requirements, the European market represents only about 10 per cent[330] of global soybean trade volume so that there are still sufficient given alternatives.[331] As we have seen in Chapter 3.5 about commodity trade flows, the North American, Middle East, North African or

[325] "For COCERAL and FEDIOL the approval of seven Voluntrary Schemes constitutes only a preliminary step towards having a harmonized internal market for sustainable biofuel raw material – COCERAL&FEDIOL Press Release - link
[326] FEDIOL, 2012.
[327] Vasquez, 2012.
[328] Passerini, 2011.
[329] Zylla, 2012.
[330] Noble Group Research, 2012b.
[331] Gallo, 2011.

Asian market among others, do not require yet any sustainability certificates. Europe takes the lead in this matter and becomes a role model for other markets.[332] Henceforward, there will be a certain time lag for the other countries to catch up so that the commodity market that has become a platform trading standardized products will slightly change to one offering sustainable certified and non-certified products. It is so far unclear how much pressure the market can cause to expand the sustainability schemes globally and will depend, as aforementioned, on political consensus building. While the certification schemes should be comprehensive and reliable as they aim to be in Europe,[333] they should also consider the differences of each region. Therefore elaborated schemes and sustainability audits should adapt to local requirements so that these criteria do not pose too much complications for producers and trading companies.[334] Over the next decade, maybe one or two of the seven already existing and recognized certificates will achieve hegemonic presence and become a uniform standard worldwide.[335] Such an emergence of an international certification body for biomass and biofuels would be feasible; nonetheless, transparency and education would be of utmost importance to guarantee both, on the supply side,[336] independency from actors along the biofuels supply chain and public authorities as well as on the demand side, consumer's recognition created through strong public information[337] and educative initiatives by the private sector.

4.5. An outlook into second and third generation biofuels

The first generation of biofuels is likely to have a time span of at least ten years, before the second generation starts becoming commercially available. Those biofuels are expected to have increased energy balances, higher efficiency levels and require less land resources leading to lower carbon footprints.[338] Advanced biofuels generally include all next generation technologies and feedstock that go beyond the current use of conventional agricultural product based such as conventional sugarcane, starch, oilseeds as well as animal fat-based biofuels (as in

[332] Noble Group Research, 2012b. and LMC International, 2009.
[333] Worldwatch Institute, 2007.
[334] Hausmann and Wagner, 2009..
[335] Kabath, 2012.
[336] Montenegro, 2011.
[337] Worldwatch Institute, 2007.
[338] Dufey, 2006

Figure 48).[339] These newer technologies aim to create an advanced biofuel derived from non-food energy crops like switch grass, mescantus or jatropha as well as agricultural, forestry and municipal waste such as used cooking oil and therefore are not in direct competition with food and feed supply.[340] The third generation moves even further as being made of algae or seaweed that does not even compete with land surfaces and does not cause land-use change effects as outlined in Chapter 3.9.[341] No commercial use of second or third generation biofuels has started yet, even though that market experts estimate the second and even third generation to reach competitive prices regarding production and market supply by 2020.[342] As biofuels are generally a political sensitive topic, not only significant investment will determine such a pathway and time-to-market, but also the upcoming legislative action being taken on a global stage.[343] Thus, current obstacles for commercial availability are expensive and time-consuming processes.[344] The on-going research and development of second-generation biofuels is putting a time frame on the current biofuel options as prediction are difficult to make about how the second generation will affect markets As economic costs for producing second generation biofuels are expected to drop faster than for biodiesel due to learning curves and second mover advantages, the technological development is supposed to proceed on an accelerated level.[345] Therefore advanced biofuels will be likely to obtain larger market shares in the upcoming decades.[346] Due to their high GHG emission saving values, advanced generation biofuels are double counted regarding their contribution to the 20:20:20 targets. Thus, current investment and subsidies will hopefully accelerate and ratios like for instance the conventional fuel vs. renewable fuel energy investments made by the World Bang Group with 49 per cent vs. 15 per cent accordingly during 2007-2009 will inverse.[347]

339 USDA FAS, 2011.
340 UFOP, 2011.
341 IEA, 2011.
342 Vasquez, 2012.
343 Worldwatch Institute, 2009.
344 IEA, 2011.
345 Eisentraut, 2010.
346 UFOP, 2011.
347 Stakeholfer Forum, 2011.

Figure 48: Commercialisation status of advanced generation biofuel technologies[348]

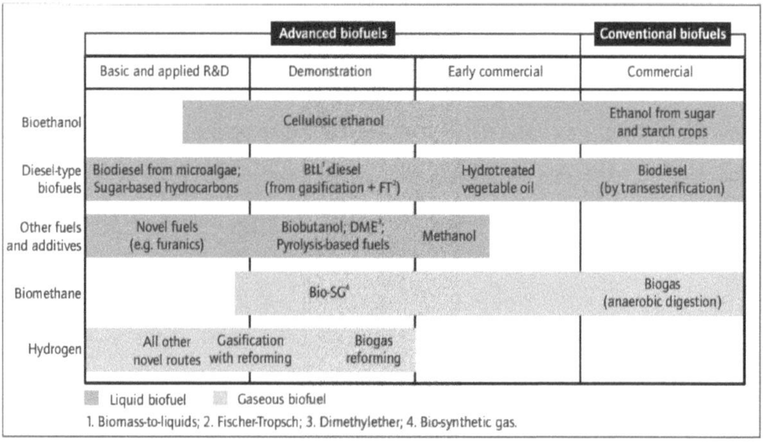

4.6. Price increases and price volatility cause food shortage and hunger

The list for sustainability criteria could be extended further. Additional sustainability criteria for the production of liquid biofuels could be soil erosion, fresh water depletion, water management, nutrient leaching or depletion, pollution prevention regarding pollutants from chemicals, impact on biodiversity, direct or indirect employment creation, minimum wages and no child labour, as well as societal responsibilities including education and health care.[349] Biofuel production from food crops, along with rising food demand in emerging countries is often associated and partly even made responsible for food price increases. Generally, these high price levels are a threat to the urban poor and to non-self sufficient farmers; nonetheless, it may also be an opportunity for farmers in the South who might benefit from higher producer prices.[350]

[348] IEA, 2011; **Biobutanol** can be used as a fuel in an internal combustion engine. It has a greater energy density and is more similar to gasoline than ethanol, and could thus be distributed through existing gasoline infrastructure. Biobutanol can be produced by fermentation of sugar via the acetone-butanol-ethanol process using bacteria such as Clastridium. **Dimethylether** (DME) can be produced from methanol through the process of catalytic dehydration or it can be produced from syngas through gasifying lignocellulosic and other biomass feedstock. Production of DME from gasification of biomass is in the demonstration stage, and the first plant started production in September 2010 in Sweden. DME is the simplest ether and can substitut for propane in liquefied petroleum gas (LPG) used as fuel, and it is a promising fuel in diesel engines, due to its high cetane number.

[349] Smeets 2004.

[350] Simantob, 2011.

The development of prices on agricultural commodity markets is fundamentally driven by supply and demand. Large funds and speculative position takers are often claimed to cause price movements and increased volatility.[351] Nevertheless, these actors provide the necessary liquidity to hedge physical positions of farmers, traders and supply chain managers. Thus, the increasing number of participants on commodity future exchanges became a necessary component of volatile commodity markets.[352] Biofuels in combination with other important factors cause price movements and are hardly responsible for resulting volatility as their demand is generally stable and less elastic to short-term price movements. The individual influence is difficult to evaluate as other forces play a much bigger role, including the weather, income and population growth, food habits, purchasing behaviour, technical progress as well as macroeconomic environment and trade-policies. For instance, the changes of food habits in emerging economies can be exemplified by the increased demand of meat in China.[353] While a Chinese consumer ate 20 kg meat during the year of 1985, today his meat consumption has increased to over 50kg, taking into consideration that on average eight kg of grain need to be fed in order to produce one kg of beef.[354] Even if there were a certain correlation between price speculation in relationship with biofuels and price volatility on agricultural commodity markets, domestic markets within developing countries would not be directly affected. Therefore, biofuels would not be directly linked to hunger and poverty. Price increases and price volatility can indeed selectively exacerbate food insecurity, but would neither be a direct cause nor significantly responsible for hunger and poverty. The food distribution problem cannot be alleviated by state intervention in order to stabilize price levels and volatility, as prices reflect only the supply-demand tightness on a macroeconomic level. Thus, price volatility cannot be avoided on inelastic food markets on a short term basis. Nonetheless, on a long term basis, the volatility of agricultural commodity prices can be reduced by infrastructure changes, open borders, free trade and enhanced market transparency.[355]

---

[351] Robles, 2009.
[352] Trostle, 2008.
[353] FAO, 2009.
[354] World Food Programme, 2009.
[355] FAO, 2011.

Instead of biofuels and price speculation, especially in developing countries, hunger and poverty are almost exclusively a result of domestic market failures caused by poor governance, corruption, civil wars, discrimination against the agriculture sector and/or triggered by extreme weather events such as droughts and floods. Furthermore, commodity price levels as these are mainly affected by fundamental shifts and development in supply and demand structures including energy prices, technological development, income and population growth as well as general economic development.[356] Nevertheless, with global shifts such as a growing population, rising living standards in economies of transition like India and China, increasing interconnectedness through mobility and global transport, that is, transport fuel consumption, there is reason to assume that land will always be scarce and the competition between new crops for energy purposes and food will remain.[357]

Effective ways to eradicate these social injustices comprise investments in agricultural production and market infrastructure, and the creation of a social safety net for the poorest. In particular, the agriculture would be exempt from commercial, agricultural and monetary burdens in order to set appropriate incentives for production and lift small farmers out of poverty integrating them into local markets activity. The industrialized countries could contribute via the reduction of trade barriers, opening their markets for developing countries as well as decrease trade distortions created by heavy subsidies to uncompetitive agricultural sectors and therefore generate at the same time a larger price-stabilizing buffer for market shocks. Finally, biofuels and their promotion should not be blamed for hunger and poverty in the world. Even their small influence on price movements on world market is quite limited and only in combination with other drivers relevant in the short run.[358] Nevertheless, the image of the biofuels industry could be significantly improved if future government support were designed with enhanced focus on transparency and sustainability. Technological progress and learning curves may help to independently operate in the future without political support schemes and regulation excesses. Only then an honest

[356] European Parliament and Council, 2009.
[357] Von Braun, 2007 and Wiesenthal, 2009.
[358] FAO, IFAD, OECD, UNCTAD, WFP, World Bank, WTO, IFPRI and UN HLTF, 2011.

cost-benefit balance might be taken for the promotion of biofuels and an objectification of current discussions can be achieved.[359] In addition, higher prices can stimulate investments in the agriculture sector; and therefore, contribute to long term improvements even in relationship with poverty and food security.[360] Since 80 per cent of the poor live in rural areas, these private sector investments could significantly contribute to sustainable development.

Besides, changes in price levels and price volatility on international markets are not directly correlated with domestic prices, as governments often introduce or maintain their own trade and pricing policies. Additionally, elevated transport costs or complete lack of infrastructure limit the access to global markets. The price transmission elasticity[361] is often much smaller than one and sometimes for remote rural areas in developing countries, even zero.[362] In case that people in poor countries do not depend on imports and eat food that is produced domestically, they are hardly affected by global commodity flows at all. In addition to price changes, domestic production fluctuations are often stronger than the fluctuations in the aggregate world production and in the case of closed markets there is barely any buffer for compensation. Lowering trade barriers and increased market integration would therefore create a stabilizing effect for households in developing countries.[363] Thus, biofuels production may have not caused increased volatility in food prices in the long term but can give price signals to farmers what kind of crop is more economically convenient to cultivate.[364] Particularly in the case of crop rotation decisions for example when a farmer has to decide between corn and soybeans, he usually gets a good orientation by the soybean-corn Chicago Board of Trade (CBOT) future price spread.

Finally, other important areas exist which are more relevant for a sustainable food access. Demographic shifts require a scaling up of humanitarian assistance capacity in order to cope with the rising number of people needing help due to droughts, floods and storms. Their operating contexts move from crisis re-

---

359 Schmitz and Moleva, 2012.
360 See FAO, 2011.
361 Domestic price change in per cent as a result of a one per cent price change on world markets.
362 Minot, 2011.
363 OECD, 2011
364 EuropaBio, 2008.

sponse to crisis prevention as for instance the mainstreaming of disaster risk reduction throughout national development programmes. Additionally, social protection systems as key element of climate change adaptation policy so that the most vulnerable people have a chance to cope with climate change and other related risks. Nicholas Stern[365] also agrees that such a social protection system offers a better approach to manage future food price spikes than less targeted measures such as price controls. These would reduce the incentives for farmers to produce more food. Moreover, economy-wide subsidies are more of a burden on public sector budgets and contribute to inflationary tendencies.[366] Besides, the focus on food production does not only refer to more food production as 50 per cent more quantity is needed by 2030 according to a World Bank, but also to the quality of food, being both more nutrient and more resilient, sustainable and equitable as 80 per cent of the world's poor, as aforementioned, live in rural areas. In addition to humanitarian assistance, there is a lack of investment in infrastructure in terms of knowledge and innovation, particularly in relationship with water and land assets since both resources are projected to become scarcer.[367] As natural resource governances are often highly politicized, the need for governments and donors to build institutional capacity in natural resource governance emerges in order to strengthen the poor and unprivileged ones. Furthermore, infrastructure investments also refer to communication technologies allowing small farmers selling their crops to larger exporters and corporations as well as the access to credit, which will be essential in helping farmers to acquire new technologies and enhance efficiency among other reasons.[368]

As actions to mitigate climate change by reducing greenhouse gas emissions would substantially reduce the risk of hunger. Stabilizing GHG concentrations at 500 parts per million (ppm) could reduce the increase by 50 per cent. This illustrates the urgent need for substantial international action and demonstrates that biofuels independently of which generation, are truly connected to hunger and

[365] Stern, 2008.
[366] United Nations High Level Task Force on the Global Food Crisis. 2008.
[367] McKinsey Global Institute, 2011.
[368] World Food Programme, 2009.

poverty, but in a positive way as they can be part of a more complex climate change solution strategy.[369]

[369] Stern, 2010.

5. Conclusion

Biofuel production and its associated global trade provide vast opportunities for developing and developed countries including economic, environmental and social benefits. The introduction of the RED as a pioneering political framework for biofuels has led to a change of mind-set within the biofuel industry in order to fight climate change. All individual members within the supply chain have become either more attentive to sustainability practices or at least more conscious about direct and indirect social and environmental implications. Nevertheless, such a fundamental change as the complete implication of the RED tends to take time and therefore the upcoming development of the EU ten-year action plan primarily needs to be observed with the implementation of a deadline in order to draw objectively a conclusion about the overall impact. Since implementation in 2009, the RED has transformed the European biofuel landscape in the following ways: increased installation capacity subsequently overall production, higher volume of imports and increased employment and job creation. The continuously rising demand and economic growth within the industry has been incentivised through mandates and politically supported mid term objectives such as the 10% biofuel share within the transportation sector. Nevertheless, the RED still leaves some areas for improvement including transparency, communication and uniformity. It is of utmost importance to enhance the transparency about the overall objective in terms of GHG savings; these should be based on solid fundamentals, such as scientific studies about the direct and indirect impacts of individual biofuels. Improved education of the public regarding different biofuel sources and the increase of available information is likely to lead to higher awareness and overall support as well as increasing the image of the members of the supply chain. Given a sound basis of fundamental support and clear objective communication, the current certification could be more uniform, possibly with one or two standards or voluntary schemes that reach a hegemonic position regionally or even globally. In addition to the RED, governments should not only introduce political frameworks but also work on more intensive tracking and monitoring of interim objectives, possibly with the use of key performance indicators (KPI's) as well as the introduction of a respected sanction scheme so that individual MS and countries are sufficiently attracted to commit to the action plan.

Finally, the RED is only a minor part of the public intervention to counteract market failures like climate change. European governments' responsibilities include the provision of long-term targets and support policies, which consider all different stakeholders that are directly and/or indirectly involved in the biofuel supply chain. Amongst others, these policies should stimulate capital and financial investments in sustainable biofuel production so that advanced generation biofuels can achieve a mature stage of commercialisation. Furthermore, increased research and development funding should be provided in order to adopt progress along the learning curve in terms of cost and efficiency gains for current generation of biofuels. In order to provide strong signals to the market, remove uncertainty, create confidence among business investors and market participants, MS should increase mandates and interim objectives that are in line with the 20:20:20 Energy Strategy targets, providing price stability on commodity markets and therefore encouraging investments. Governments could also set a precedent by consecutively lowering subsidies to conventional (fossil) fuels so that a realistic price for CO2 emission and a value for GHG emissions savings can be developed.

In addition to public sector initiatives, industrial companies along the supply chain should engage in the development of independent and uniform sustainability schemes that create a globally recognised standard for biofuels. Public-private partnerships could create incentives for smaller companies to take an active part within the biofuel industry leading to sustainable development particularly in rural areas within developing countries. It is critical that small-scale biofuel development occurs alongside larger scale production for export. Policies should be oriented towards potential benefits for both large and small-scale production and should support the most environmentally and socially beneficial production processes available, always considering the population and economic growth that tends to lead to a rising demand. Technology transfer and investment from industrialised countries to facilitate market development remains elementary for further development regarding advanced biofuels. In addition to the stakeholders directly affected by the biofuel value chain, research institutes and the international science community should further enhance LCA for biofuels so that an overall understanding of the direct and indirect effects of land use change among others so it can be quantified. Non-Government Organisations (NGOs) should keep an eye on the industry and government parties so that continuous monitoring by third parties can be guaranteed. Furthermore, NGOs can contribute substantially in terms of education of the final consumer by providing objective information about certifications and operational practices. As outlined in Chapter 4, combining mandates, government purchasing policies and the supportive interventions for biofuel-compatible infrastructure and technologies have been proven to be most successful in developing reliable biofuel markets. Other policies including financial initiatives, such as loan guarantees, consumption and manufacturer tax incentives, as well indirect subsidies will fade out as technologies become commercially viable and self sustainable. Future policies should be based on the RED and further refer to climate change related areas, explaining the scope of biofuels including land use, transport, energy, and housing as well as education and information programmes to raise awareness of the impact of consumption patterns and finally leading to a consecutive removal of subsidies that promote unsustainable patterns of production and consumption.

The RED could set the course for secure, competitive and low carbon energy in the EU27 by 2020. Nevertheless, this biofuel roadmap requires other governments to join the European renewable energy movement and adapt their sustainability standards in order to have a true impact on climate change on a global scale. The European market in comparison with the potential for future demand coming from China and India is relatively small. As indicated in Figure 49, the Western world and especially Europe remains the dominant player that have an important role to play currently within the biofuel marketplace but by 2050 it is expected to become less relevant representing less than 50 per cent of global demand.

Figure 49: Biofuel demand forecast by region 2010-2050[370]

[370] IEA, 2010. FSU refers to Former Soviet Union.

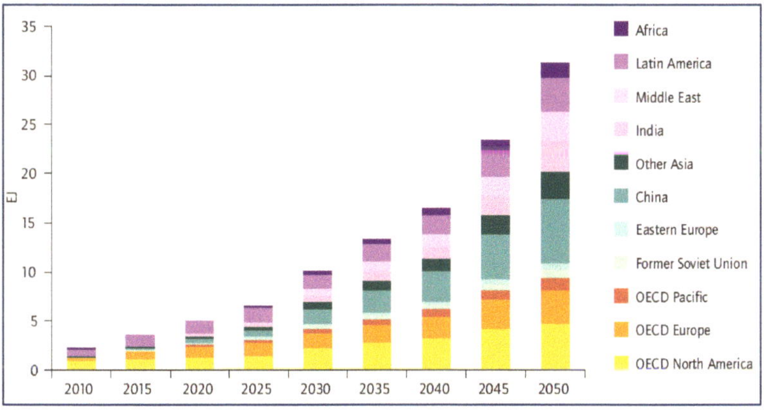

Biofuel production and its associated global trade provide vast opportunities for developing and developed countries including economic, environmental and social benefits. The introduction of the RED as a pioneering political framework for biofuels has led to a change of mind-set within the biofuel industry in order to fight climate change. All individual members within the supply chain have become either more attentive to sustainability practices or at least more conscious about direct and indirect social and environmental implications. Nevertheless, such a fundamental change as the complete implication of the RED tends to take time and therefore the upcoming development of the EU ten-year action plan primarily needs to be observed with the implementation of a deadline in order to draw objectively a conclusion about the overall impact. Since implementation in 2009, the RED has transformed the European biofuel landscape in the following ways: increased installation capacity subsequently overall production, higher volume of imports and increased employment and job creation. The continuously rising demand and economic growth within the industry has been incentivised through mandates and politically supported mid term objectives such as the 10% biofuel share within the transportation sector. Nevertheless, the RED still leaves some areas for improvement including transparency, communication and uniformity. It is of utmost importance to enhance the transparency about the overall objective in terms of GHG savings; these should be based on solid fundamentals, such as scientific studies about the direct and indirect impacts of individual biofuels. Improved education of the public regarding different biofuel sources and the increase of available information is likely to lead to higher awareness and overall support as well as increasing the image of the members of the supply chain. Given a sound basis of fundamental support and clear objective communication, the current certification could be more uniform, possibly with one or two standards or voluntary schemes that reach a hegemonic position regionally or even globally. In addition to the RED, governments should not only introduce political frameworks but also work on more intensive tracking and monitoring of interim objectives, possibly with the use of key performance indicators (KPI's) as well as the introduction of a respected sanction scheme so that individual MS and countries are sufficiently attracted to commit to the action plan.

Figure 50: Global GHG abatement cost curve beyond business-as-usual (2030)[371]

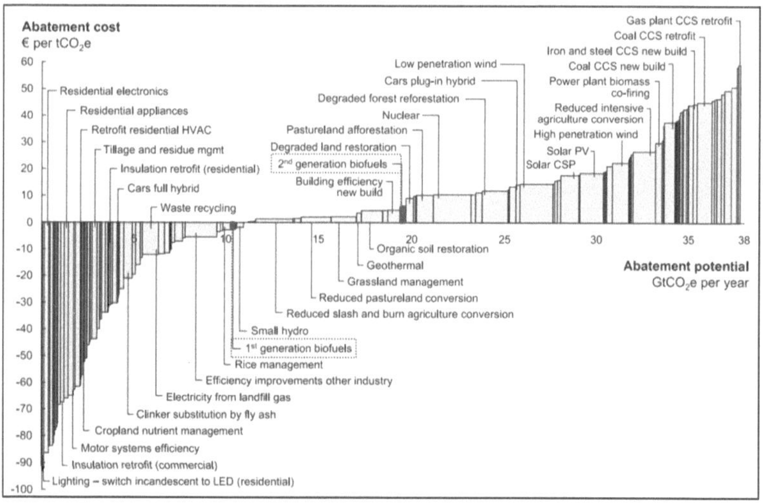

"The oil will never run out and it is not of it because we have a lot of it. The Stone End did not end because of a lack of stones and the Oil age will certainly end long before the world runs out of oil. Finally, it is ideas, innovation and technologies that end the Age of Oil, long before we will run out of oil"[372]

[371] The curve presents an estimate of the maximum potential of all technical GHG abatement measures below 60€ per tCO2e if each lever was pursued aggressively. It was not developed as forecast of what role different abatement measures and technologies will play. McKinsey & Company, 2009.
[372] Nussbaum, 2011, Stevens, 2008, originally sheikh Zaki Yamani, a Saudi Arabian oil minister used that quote in the 70s.

6. Bibliography

Abengoa (2012): "Abengoa Press release on biofuels", http://www.abengoa.es/htmlsites/boletines/en/ abril2011/bioenergy.html, *accessed online on January, 14th 2012.*

ABIOVE (2012): Brazilian Biodiesel Newsletter, www.biodieselbr.com, accessed on January, 10th 2012

Accenture (2007): "Irrational Exuberance"? An assessment of how the burgeoning biofuels market can enable high performance, http://www.accenture.com/SiteCollectionDocuments/PDF/Accenture _Irrational_Exuberance.pdf#zoom=50, *accessed online on January, 10th 2012.*

ADM (2012): company website, www.adm.com, *accessed online on January, 12th 2012.*

European Biomass Association (AEBIOM, 2012): "Annual Statistical Report 2011", http://www.aebiom.org/?cat=20, *accessed online on January, 8th 2012.*

AEE (2009): Arbeitsgemeinschaft Erneuerbarer Energien, figures, http://www.aee.at/, *accessed online on January, 8th 2012.*

Agroparlamento, (2012): Organisation Website, Impuestos de exportacion de differntes granos, http://www.agroparlamento.com/agroparlamento/default.asphttp://www.agroparlamento.co m/agroparlamento/default.asp, , *accessed online on January, 8th 2012.*

Almada, M. (2006): "Perspectivas de los biocombustibles en la Argentina", Seminario Agroenergía – Biocombustibles, Santiago de Chile. http://www.cne.cl/hidrocarburos/f_biocombustibles.html, *accessed online on December, 10th 2011.*

Alcamo, J. (2009): Chief Scientist, UNEP, Intergovernmental Panel on Climate Change (IPCC), in the Summary Report B4E Climate Summit 2009, Paris, P. 5.

Agência Nacional do Petróleo (ANP, 2011): "Gás Natural e Biocombustíveis - Produção de Biodiesel" - Metros Cúbicos, Sao Paulo, Brazil.

Asal, S. et al (2006): "Opportunities for and obstacles to sustainable biodiesel production in Argentina", Energy for Sustainable Development, Volume X, No. 2, June, pp. 48-58

Bacon, R. (2001): "Petroleum Taxes." Viewpoint Note 240, http://rru.worldbank.org/Documents/PublicPolicyJournal/240Bacon-831.pdf, *accessed online on January, 10th 2012.*

Bauen, A., et al. (2009): "Bioenergy - a sustainable and reliable energy source. A review of status and prospects"., IEA Bioenergy: ExCo:2009:06, http://www.task39.org/LinkClick.aspx?fileticket =8IsypIOAwXs%3D, *accessed online on January, 10th 2012.*

Beyond Petrol (BP, 2011): "Energy Outlook 2030", corporate reports and publications, http://www.bp.com/liveassets/bp_internet/globalbp/globalbp_uk_english/reports_and_publi cati- ons/statistical_energy_review_2011/STAGING/local_assets/pdf/2030_energy_outlook_boo klet.pdf, *accessed online on January, 10th 2012.*

BioDieselBr.com (2011): "Conferencia 2011 – Um olhar global sobre o biodiesel", 31.10.2011, www.biodieselbr.com, *accessed online on January, 10th 2012.*

Biomass Biofuel Sustainability voluntary Scheme (2BSvs, 2012): http://en.2bsvs.org/, *accessed online on November 8th 2012.*

Biomass Research and Development Initiative (BRDI, 2008)**: "**The Economics of Biomass Feedstocks in the United States", http://www.osti.gov/bridge/purl.cover.jsp?purl=/219271-innoc8/webviewable/219271.pdf, *accessed online on January, 10th 2012.*

Bonsucro (2012): company website, http://www.bonsucro.com/certification.html, *accessed online on January, 10th 2012.*

Bundesministerium für Ernährung, Landwirtschaft und Verbraucherschutz (BELV, 2012): „Statistischer Monatsberichte", http://www.bmelv-statistik.de/de/statistischer-monatsbericht, accessed *March, 12th 2012.*

Bundesministerium fuer Umwelt , Naturschutz und Reaktorsicherheit (BMU, 2012): "Entwicklung der erneuerbaren Energien in Deutschland im Jahr 2011, http://www.erneuerbare-energien.de/erneuerbare_energien/datenservice/schaubilder/doc/42038.php, accessed *March, 12th 2012.*

Bureau Veritas (2012): Organisation Website, http://www.bureauveritas.de/wps/wcm/connect/bv_de/ Local/Home/About-Us/Profile-Logo/, accessed *March, 12th 2012.*

Buringh, P. and Dudal, R. (1987) "Agricultural Land Use in Space and Time". In: M.G. Wolman and F.G.A. Fournier (eds.) Land Transformation in Agriculture. New York, Wiley.

Camara Argentina de Biocombustivels (CARBIO, 2012): Organization Website, http://www.carbio.com.ar/en/?con=sustentabilidad, *accessed on March 10h, 2012.*

Clarke, R. and King, J. (2004): "The Atlas of Water", London, Earthscan Books.

Comité du Commerce des céréales, aliments du bétail, oléagineux, huile d'olive, huiles et graisses et agrofournitures Coceral (2012): association website, http://www.coceral.com/

Conservation International (CI, 2012): Organisation Website, http://www.biodiversityhotspots.org/xp/hotspots/Pages/default.aspx, *accessed on February, 17th 2012.*

Consumers International (2012): "Rio+20 Explained", The Global Voice for Consumers, http://www.consumersinternational.org/our-work/sustainable-development/key-projects/at-the-un/earth-summit-2012-(rioplus20), *accessed online on January, 17th 2012.*

Daab, S. (2012): Interview with Sascha Daab, Commodity Trader at Goldman Sachs.

Demirbas, A. (2008): "Biodiesel: A Realistic Fuel Alternative for Diesel Engines", http://books.google.no/books/about/Biodiesel.html?id=0vBalrSH_OEC&redir_esc=y, *accessed online on February, 26th 2012.*

De Souza, A. (2011): Interview with Adilson de Souza, Director of Noble Fertilizantes Brazil in Sao Paulo on November, 24th 2011.

Delzeit, R. et al. (2011): "Assessing the Land Use Change Consequences of European Biofuel policies and its uncertainties", Kiel Institute of the World Economy, Study on the behalf of the European Biodiesel Board, http://www.ebb-eu.org/EBBpressreleases/Review_iLUC_IfW_final.pdf, *accessed online on February, 26th 2012.*

Deutsches Biomasseforschungszentrum gGmbH (DBFZ, 2011): "Monitoring zur Wirkung des Ernerubare-Energien-Gesetzes (EEG) auf die Entwicklung der Stromerzeugung aus Biomasse", Dritter Zwishcenbericht, Leipzig.

Dobbs, Richard et. al. (2011): "Resource Revolution: Meeting the world's energy, materials, food, and water needs", McKisney Global institute, http://www.mckinsey.com/Insights/MGI/Research/ Natural_Resources/Resource_revolution, *accessed online on February, 26th 2012.*

Doornbosch, R. and Steenblik, R. (2007): "Biofuels: is the cure worse than the disease?" SG/SD/RT(2007)3, OECD, Paris, http://www.oecd.org/dataoecd/15/46/39348696.pdf, *accessed online on February, 26th 2012.*

Dufey, A. (2006): "Biofuels production, trade sustainable development: emerging issues", Environmental Economics Programme/Sustainable Markets Group, International Institute for Environment and Development, http://pubs.iied.org/pdfs/15504IIED.pdf, *accessed online on February, 26th 2012.*

Editora Gazeta Santa Cruz Ltda (2011): „Annuario Brasileiro da Soja 2011" http://www.gaz.com.br/tratadas/eo_edicao/27/2011/09/20110921_a98e21950/pdf/2922_soja2011internet.pdf, *accessed online on December, 04th 2011..*

Ebrahimi, H. (2009): "IEA warns of oil 'supply crunch'". Daily Telegraph, 16 February 2009

Edenhofer. O. et al. (2011): "IPCC Special Report on Renewable Energy Sources and Climate Change Mitigation", Cambridge University Press, Cambridge, UK and New York, http://srren.ipcc-wg3.de/report/IPCC_SRREN_Full_Report.pdf, *accessed online on February, 26th 2012..*

Eisentraut, A. (2010): "Sustainable Production of Second Generation Biofuels", International Energy Agency, http://www.iea.org/papers/2010/second_generation_biofuels.pdf, *accessed on January, 13th 2012.*

Elgowainy, A. et al. (2009): "Well-to-Wheels Energy Use and Greenhouse Gas Emissions Analysis of Plug-in Hybrid Electric Vehicles", Energy Systems Division, ANL/ESD/09-2, U.S. Department of Energy, Office of Scientific and Technical Information, http://www.transportation. anl.gov/pdfs/TA/559.pdf, *accessed on January, 13th 2012.*

ENERS Energy Concept (2010), "Biofuels Platform 2010" , http://www.google.de/url?sa=t&rct=j&q=eners%20energy%20concept%2C%202010&source=web&cd=2&ved=0CDIQFjAB&url=http%3A%2F%2Fwww.biofuels-platform.ch%2Fmedias%2Fdownload.php%3Fget%3D433&ei=S8eVT8DLJ-bY4QSDxI3RDw&usg=AFQjCNFeqwQNbTc4lJF2-U8EwKGknMWrXw, *accessed on January, 13th 2012.*

Energy Sector Management Assistance Programme (ESMAP, 2006): "Annual Report 2005", http://www.esmap.org/esmap/sites/esmap.org/files/AR_2005.pdf, *accessed on April, 24th 2012.*

EurObserv'Er (2011a): "Barometre Biocarburants, http://www.eurobserv-er.org/pdf/biofuels_2011.pdf, *accessed on April, 13th 2012.*

EurObserv'Er (2011b) "Biofuels Barometer", several issues, Observ'er, IJS, ECN, Eclarean.

EuropaBio (2008): "EuropaBio Fact Sheet", Biofuels and developing countries, http://www.europabio.org/Biofuels/Biofuels_about.htm, *accessed on January, 18th 2012.*

European Biodiesel Board (EBB, 2008) "An economic and security of supply analysis of the widening EU diesel deficit". Brussels, Belgium.

European Biodiesel Board (EBB, 2010): Biodiesel statistics, http://www.ebb-eu.org/stats.php, *accessed on April, 8th 2012.*

European Biodiesel Board (EBB, 2011): "Commission consultation on accounting methods and conditions for the 10 per cent renewable energy in transport target and on the need for additional types of biofuels being listed in Annex III of the Renewable Energy Directive", Brussels http://www.envforum.eu/doku.php/news/european_commission_opens_public_consultation_on_accounting_methods_and_conditions_for_the_10_per_cent_renewable_energy_in_transport_target, *accessed on April, 13th 2012.*

European Bioethanol Fuel Association (EBIO, 2010): Ethanol statistics. Available at: http://www.ebio.org. *(Since December 2010 the EBIO webpage changed to the European Renewable Ethanol Association - ePURE, http://www.epure.org/, accessed on April, 13th 2012.*

European Commission (EC, 2010): "Energy 2020, A strategy for competitive, sustainable and secure energy", http://ec.europa.eu/energy/strategies/2010/2020_en.htm, *accessed on January, 13th 2012.*

European Commission (EC, 2009): "The Renewable Energy Progress Report: Commission Report in accordance with Article 3 of Directive 2001/77/EC, Article 4(2) of Directive 2003/30/EC and on the implementation of the EU Biomass Action Plan", COM(2005)628, Brussels.

European Commission (EC, 2006): "Biofuels in the European Union - A vision for 2030 and beyond", Final report of the Biofuels Research Advisory Council, Directorate General for Research, Brussels, http://ec.europa.eu/research/energy/pdf/biofuels_vision_2030_en.pdf *accessed on February, 13th 2012.*

European Commission Regulation (2009): "No 194/2009 of 11 March 2009 imposing a provisional countervailing duty on imports of biodiesel originating in the United States of America" Brussels: European Commission; http://www.ebb-eu.org/EBBpressreleases/AD%20Regulation%20countervailing %2020090311.pdf, *accessed online on February, 26th 2012.*

European Parliament and Council (2009): "Directive 2009/28/EC" on the promotion of the use of energy from renewable sources and amending and subsequently repealing Directives 2001/77/EC and 2003/30/EC, Brussels.

European Parliament and Council (2003): "Directive 2003/30/EC" on the promotion of the use of biofuels or other renewable fuels for transport, Brussels.

European Parliament and Council (2001): "Directive 2001/77/EC" on the promotion of electricity produced from renewable energy sources in the internal electricity market, Brussels.

European Renewable Energy Council (EREC, 2008): Renewable Energy - Technoloy Roadmap, http://www.erec.org/fileadmin/erec_docs/Documents/Publications/Renewable_Energy_Technology_Roadmap.pdf, *accessed on February, 16th 2012.*

European Union (EU, 2010): "Impacts of the EU biofuel target on agricultural markets and land use: a comparative modelling assessment", Luxembourg: Publications Office of the European Union, IPTS gro-Economic Modelling Platform, http://ec.europa.eu/energy/renewables/studies/doc/ land_use_change/study_jrc_biofuel_target_iluc.pdf, *accessed on March, 05th 2012.*

European Voice (2011): Commission to fudge CO2 effects of biofuel, http://www.europeanvoice.com/ar ticle/imported/commission-to-fudge-co2-effects-of-biofuel/72062.aspx, *accessed on March, 05th 2012.*

EUROSTAT (2012): Data Explorer, http://epp.eurostat.ec.europa.eu/portal/page/portal/eurostat/home/, *accessed on February, 16th 2012.*

EUROSTAT. (2011): Data Explorer - EU27 Trade since 1995 by CN8. Brussels, http://epp.eurostat. ec.europa.eu/, *accessed on February, 16th 2012.*

Fachagentur Nachwachsender Rohstoffe e.V. (2012): „Grünes Licht für erste EU-Systeme zum Nachweis der Nachhaltigkeit von Biokraftstoffen", http://www.bioenergie-portal.info/thueringen/ news/news/archive/2011/august/article/gruenes-licht-fuer-erste-eu-systeme-zum-nachweis-der-nachhaltigkeit-von-biokraftstoffen/?tx_ttnewspercent5Bdaypercent5D=22&cHash= 0911b1d16550a7e4b9cea4daad898edf, *accessed on February, 17th 2012.*

Fachagentur Nachwachsender Rohstoffe e.V. (2011a): „Biodiesel", http://www.biokraftstoffe.info/kraftstoffe/biodieesl/, *accessed on February, 17th 2012.*

Fachagentur Nachwachsender Rohstoffe e.V. (2011b): „Biokraftstoffe", http://www.biokraftstoffe.info/kraftstoffe/, *accessed on February, 17th 2012.*

FEDIOL (2012): Website FEDIOL – Statistical Data and Press Releases, http://www.fediol.be/, *accessed on March, 05th 2012.*

Fitzgerald, S. (2012): EU2020 – Energy and Climate Change, International Institute of European Affairs, http://www.iiea.com/blogosphere/eu2020--energy-and-climate-change-infographic, *accessed on February, 16th 2012.*

Freedman, B. et al. (1984): "Affecting the Yields of Fatty Esters from Transesterified Vegetable Oils", J. Am. Oil Chem. Soc. 61: 1638– 1643, http://ddr.nal.usda.gov/bitstream/10113/19247/1/ADL85030004.pdf, *accessed on April, 20th 2012.*

FO Licht (2010): World Ethanol & Biofuels Report. Kent, UK & Ratzeburg, Germany: Informa UK Ltd. Vol. 8 (16), http://www.agra-net.com/portal2/puboptions.jsp?Option=archive&pubid=ag072, *accessed on February, 19th 2012.*

Food and Agriculture Organization of the United Nations (FAO, 2012): FaoStat, http://faostat3.fao.org/home/index.html, . *accessed on February, 19th 2012.*

Food and Agriculture Organization of the United Nations (FAO, 2011): The state of food insecurity in the world. How does international price volatility affect domestic economies and food security? FAO of UN, Rome, 2011.

Food and Agriculture Organization of the United Nations (FAO, 2009): "1.02 billion people hungry", News release, 19 June, http://www.fao.org/news/story/en/item/20568/icode/, *accessed on February, 17th 2012.*

Food and Agriculture Organization of the United Nations (FAO, 2008): "The State of Food and Agriculture", Rome, http://www.fao.org/docrep/011/i0100e/i0100e00.htm, *accessed on February, 17th 2012.*

Food and Agriculture Organization of the United Nations (FAO, 2006): Livestock's Long Shadow: Environmental Issues and Options. Rome, ftp://ftp.fao.org/docrep/fao/010/a0701e/a0701e.pdf, *accessed on February, 19th 2012.*

Food and Agriculture Organization of the United Nations, IFAD, OECD, UNCTAD, WFP, The World Bank, The WTO, IFPRI and UN HLTF (2011): Price volatility and agricultural markets: Policy responses. 2nd June 2011. Prof. Dr. Dr. H.C. P. Michael Schmitz Institut Für Agrarpolitik Und Marktforschung Justus-Liebig-Universität Gießen Und Institut Für Agribusiness – Vorstudie - Bestimmungsgründe Für Das Niveau Und Die Volatilität Von Agrarrohstoffpreisen Auf Internationalen Märkten – Implikationen Für Welternährung Und Politikgestaltung, Gießen, Februar 2012 im Auftrag des Verbands der Deutschen Biokraftstoffi ndustrie e. V. und der Union zur Förderung von Öl- und Proteinpfl anzen e. V.

Forest Stewardship Council (FSC, 2012): Website http://www.fsc.org/pc.html, *accessed on February, 17th 2012.*

Gallo, L. (2011): Interview with Lucas Gallo, Senior Trader and head of Argentinian Soybean Crush Activity in Buenos Aires, Argentina on December, 1st 2011.

German Union for the Promotion of Oil and Protein Plants (UFOP, 2011): "International biodiesel markets- developments in production and trade", http://www.ufop.de/downloads/EV_Ecofys-UFOP_en_2012.pdf, *accessed on January, 10th 2012.*

German Union for the Promotion of Oil and Protein Plants (UFOP, 2010): "Biodiesel 2009/2010" Berlin, http://www.ufop.de/downloads/RZ_UFOP_0879_netz_DE.pdf, *accessed on January, 10th 2012.*

German Union for the Promotion of Oil and Protein Plants (UFOP, 2010): "Biodiesel 2010/2011" Berlin, http://www.ufop.de/downloads/Auszug_Biodiesel_D_2011_web.pdf, *accessed on January, 10th 2012.*

Global Renewable Fuels Alliance (2011): "Global Ethanol Production to Reach 88.7 Billion Litres in 2011", World Ethanol Production Playing an Increasing Role in Energy Security, http://www.marketwire.com/press-release/global-ethanol-production-to-reach-887-billion-litres-in-2011-1395312.htm, *accessed on February, 17th 2012.*

GMO-Compass (2012a): "Genetically modified Organism Compass database", http://www.gmo-compass.org/eng/agri_biotechnology/gmo_planting/342.genetically_modified_soybean_global_area_under_cultivation.html, *accessed on February, 17th 2012.*

GMO-Compass (2012b): based on USDA (2012): "National Agricultural Statistic Service" http://usda.mannlib.cornell.edu/MannUsda/viewDocumentInfo.do?documentID=1000, *accessed on February, 17th 2012.*

Graboski, M. (2002): "Fossil energy use in the manufacture of corn ethanol", Chesterfield, www.oregon.gov/ENERGY/RENEW/Biomass/docs/FORUM/FossilEnergyUse.pdf?ga=t, *accessed on February, 17th 2012.*

GRAIN (2007a): "Stop the agrofuel craze!", http://www.grain.org/article/entries/597-stop-the-agrofuel-craze, *accessed on February, 17th 2012.*

GRAIN (2007b): "Corporate power: Agrofuels and the expansion of agribusiness", http://www.grain. org/article/entries/598-corporate-power-agrofuels-and-the-expansion-of-agribusiness, *accessed on February, 17th 2012.*

GRAIN (2007c): "Further reading: good materials on agrofuels", http://www.grain.org/ article/entries/608-further-reading-good-materials-on-agrofuels, *accessed on February, 17th 2012.*

GRAIN (2007d): "Latin America - Soya nexus in South America", http://www.grain.org/ article/entries/619-latin-america-soya-nexus-in-south-america, *accessed on February, 17th 2012.*

Greenea (2011): "Market Analysis Greenea", Market report for Used Cooking Oils in Europe, Founder, Fabien Hillairet, fabien.hillairet@greenea.com

Greenergy (2012): corporate Website, http://www.greenergy.com/, *accessed on February, 17th 2012.*

Greenpeace (2011): "Feedstock used for Biodiesel 2011, Europe", based on "Greenpeace investigation on diesel, July 2011", http://www.greenpeace.de/fileadmin/gpd/ user_upload/themen/klima/Test_Biodiesel_11_-_Europa.pdf, *accessed on April, 24th 2012.*

Greenpeace (2007): "Cooking the Climate – How the Palm Oil Industry is", http://www.greenpeace.org/france/PageFiles/266591/cooking-the-climate.pdf, *accessed on March 10h, 2012.*

Guerrero, L. (2012): Interview with Laurent Guerrero, Head of Global Trading of Grains and Oilseeds at Noble Resources S.A., Geneva on March, 1st 2012.

Haas, R. And Remmele, Dr. E. (2011): „Status Quo der dezentralen Oelgewinnung – bundesweite Befragung", Technologie und Foerderzentrum im Kompetenzzentrum fuer Nachwachsende Rohstoffe, Berichte aus dem TFZ, Im Auftrag der Union zur Foerderung von Oel und Preoteinplanzen, http://pflanzenoel.agrarplus.at/pdf/tfz_ufop_dezentrale_oelgewinnung.pdf, *accessed online on February, 11th 2012.*

Harvard University (2007): "Implications of a Future Global Biofuels Market for Economic Development and International Trade", Report of the John F. Kennedy School of Government Workshop on Biofuels, http://www.hks.harvard.edu/var/ezp_site/storage/fckeditor/file/ pdfs/centers-programs/centers/mrcbg/programs/ssp/docs/events/workshops/2007/biofuel/ Harvard_Biofuels_Workshop_report_070509.pdf, *accessed on February, 17th 2012.*

Hausmann, R. and Wagner, R. (2009): "Certification Strategies, Industrial Development and a Global Market for Biofuels", Center for International Development, Cambridge, MA: Harvard University, http://www.hks.harvard.edu/var/ezp_site/storage/fckeditor/file/pdfs/centers-programs/centers/cid/publications/faculty/wp/174.pdf, *accessed on February, 17th 2012.*

Hill, J. et al. (2006): "Environmental, economic, and energetic costs and benefits of biodiesel and ethanol biofuels", PNAS 103(30): 11206-11210, http://www.pnas.org/content/103/30/11206.abstract, *accessed on February, 17th 2012.*

Huo, H. et al. (2008): "Life-cycle assessment of energy and greenhouse gas effects of soybean derived biodiesel and renewable fuels", ANL/ESD/08-2. Argonne, Ill.: Argonne National Laboratory. Available at: www.transportation.anl.gov/pdfs/AF/467.pdf, *accessed on January, 11th 2012.*

Institute of European Environmental Policy (IEEP, 2010): Anticipated Indirect Land Use Change Associated with Expanded Use of Biofuels and Bioliquids in the EU – An Analysis of the National Renewable Energy Action Plans, November 2010.

International Energy Agency (IEA, 2012): "Oil Market Report April Highlights"http://omrpublic.iea.org/currentissues/full.pdf, *accessed 15 May, 2011.*

International Energy Agency (IEA, 2011): "Technology Roadmap – Biofuels for Transport", http://www.iea.org/papers/2011/Biofuels_Roadmap.pdf, *accessed 15 May, 2011.*

International Energy Agency (IEA, 2010): "Energy Technology Perspectives 2010", OECD/IEA, Paris, http://www.iea.org/Textbase/nppdf/free/2010/etp2010_part1.pdf, *accessed 15 May, 2011.*

International Energy Agency (IEA, 2009): "Transport, Energy and CO*2*. Moving towards mobility", OECD/IEA, Paris, http://www.iea.org/textbase/nppdf/free/2009/transport2009.pdf, *accessed 15 May, 2011.*

International Energy Agency (IEA, 2008): "World Energy Outlook 2008", Paris, http://www.iea.org/textbase/nppdf/free/2008/weo2008.pdf, *accessed 15 May, 2011.*

International Energy Agency (IEA, 2007): "Contribution of renewables to energy security", IEA Information Paper, Energy Technology Collaboration Division, IEA, Paris, http://www.iea.org/papers/2007/so_contribution.pdf, *accessed 15 May, 2011.*

International Energy Agency (IEA, 2004): "Biofuels for Transport An International Perspective" International Energy Agency, Paris, http://www.cti2000.it/Bionett/All-2004-004%20IEA%20biofuels%20report.pdf, *accessed 15 May, 2011.*

International Food and Agricultural Trade Policy Council (IPC, 2006): "IPC Talks Doha, Cotton and Biofuels", News Release, http://www.agritrade.org/pressroom/releases/Doha_Cotton_Biofuels_Oct_06.pdf, *accessed on March 10h, 2012.*

International Monetary Fund (2008): World Economic Outlook. Washington DC.

International Sustainability and Carbon Certification (ISCC, 2012): corporate Webpage, http://www.iscc-system.org/, *accessed on February, 17th 2012.*

McCurry, J. (2007): "Analysis of Glycerin and Glycerides in Biodiesel (B100) Using ASTM D6584 and EN14105, ASTM D6571 standard of Free and Total Glycerin Specifications for Biodiesel", Wilmington, http://www.chem.agilent.com/Library/applications/5989-7269EN.pdf, *accessed on February, 17th 2012.*

Kabath, J (2012): Interview with Jan Kabath, Biofuels Trader and Head of Louis Dreyfus Commodities Biofuels Germany at Louis Dreyfus Commodities Suisse S.A., 04th March, 2012.

Kaltschmitt, M et al. (2009): „Energie aus Biomasse", Grundlagen, Techniken und Verfahren, Springer Verlag, Heidelberg, ISBN: 978-3-540-85094-6.

Ki-Moon, B. (2011): Remarks at event on "Clean Industrial Revolution", Durban – South Africa, http://www.un.org/apps/news/infocus/sgspeeches/search_full.asp?statID=1407, *accessed on February, 17th 2012.*

Knothe, G. et al. (2005): "The Biodiesel Handbook", National Center for Agricultural Utilization Research Agricultural Research Service, U.S. Department of Agriculture, http://www.bmstu.ru/ps/~lib/fileman/download/4%20%D0%BA%D1%83%D1%80%D1%81%81/The%20biodiesel%20handbook.pdf, *accessed on January, 10th 2012.*

Koplow, D. (2007): "Biofuels—At What Cost? Government support for ethanol and biodiesel in the United States: 2007 update." Geneva: Global Subsidies Initiative. www.globalsubsidies.org/files/assets/Brochure_-_US_Update.pdf, *accessed on January, 11th 2012.*

Lamers P. et al (2011): "A. International bioenergy trade – a review of past developments in the liquid biofuels market", Renewable and Sustainable Energy Reviews, Vol. 15, pages 2655-2676, http://www.sciencedirect.com/science/article/pii/S1364032111000712, *accessed on January, 11th 2012.*

Laney, L. (2006): "Biofuels: Promises and Constraints", IPC Discussion Paper, http://www.agritrade.org/Publications/DiscussionPapers/IPC_Biofuels_Promises%20and%20Constraints.pdf, *accessed on March 10h, 2012.*

Lawrence, L. (2010): "How Good Politics Results in Bad Policy: The Case of Biofuel Mandates", Center for International Development, Cambridge, MA: Harvard University, http://www.hks.harvard.edu/var/ezp_site/storage/fckeditor/file/pdfs/centers-programs/centers/cid/publications/faculty/wp/200.pdf, *accessed on February, 17th 2012.*

LMC international (2009): "Global production of biodiesel by continent", Oxford, http://www.lmc.co.uk.

Lui, D. (2012): Interview with Senior Soybean Trader Dennis Lui at Noble Group Ltd. on March, 14th and April 25th 2012.

Malthus, T. (1970): "An essay on the principle of population", New York: Penguin, 1970; originally published in 1798.

Mathews, J. (2008): "Towards a sustainably certifiable futures contract for biofuels", http://www.sciencedirect.com/science/article/pii/S0301421508000219, *accessed on February, 17th 2012.*

McCurry, J. (2007): "Analysis of Glycerine and Glycerides in Biodiesel (B100) Using ASTM D6584 and EN14105", Wilmington, USA, http://www.chem.agilent.com/Library/applications/5989-7269EN.pdf, *accessed on January, 11th 2012.*

McKinsey & Company (20009): "Pathways to a Low-Carbon Economy", Version 2 of the Global Greenhouse Gas Abatement Cost Curve, https://solutions. mckinsey.com/ClimateDesk/default.aspx, *accessed on February, 17th 2012.*

McKinsey Global Institute (2011): "Resource Revolution: Meeting the world's energy, materials, food and water needs", McKinsey sustainability & Resource Productivity Practice, http://www.mckinsey.com/Features/~/media/McKinsey/dotcom/homepage/2011%20Nov%20Resource%20Revolution/Resource_revolution_full_report_v2.ashx, *accessed on February, 17th 2012.*

Methanol Institut (MI, 2009): Physical properties of pure methanol, Alexandria, Va., www.methanol.org/TechnicalInformation/Resources/Technical-Information/Physical-Properties-of-Pure-Methanol.aspx, *accessed online on December, 12th 2011.*

Minot, N. (2011): "Transmission of World Food Price Changes to Markets in Sub-Saharan Africa", Ifpri Discussion Paper 01059, http://www.ifpri.org/publication/transmission-world-food-price-changes-markets-sub-saharan-africa, *accessed on April, 20th 2012.*

Monsanto (2012): Soybean Seeds, http://www.monsanto.com/products/Pages/soybean-seeds.aspx, *accessed on March 10h, 2012.*

Montenegro, G. (2011): Interview with Gabriela Montenegro, Head of Quality and Environment of Noble Argentina in Rosario on December, 7th 2011.

Mousdale, D. (2008): "Biofuels: Biotechnology, Chemistry, and Sustainable Development, CRC Press, Taylor and Francis Group, London.

Multifunctional Carbon Nanotubes for Biomedical Applications (2012): CARBIO Webpage, www.carbio.eu, *accessed online on February, 11th 2012.*

Mukhtar, M. and Singhal, A. (2012): "When money grows on tress - Good base case, with risks for biased to the upside; look for cheapest exposure to CPO", Nomura Equity Research, http://pg.jrj.com.cn/acc/Res/CN_RES/INVEST/2012/3/23/df00e34b-19ea-4adf-a273-06fa4e3558e5.pdf, *accessed on April, 20th 2012.*

National Grid (2011): "UK Future Energy Scenarios", UK gas and electricity transmission, http://www.nationalgrid.com/NR/rdonlyres/86C815F5-0EAD-46B5-A580-A0A516562B3E/50819/10312_1_NG_Futureenergyscenarios_WEB1.pdf, *accessed on April, 20th 2012.*

Noble Argentina (2011): Visit of the Nonble Argentina Timbues Soybean collection, preparation and crushing plant, Timbues.

Noble Group Research (2012a): Animal feed and livestock analysis Team, Geneva.

Noble Group Research (2012b): Trade Consolidation of Grains and Oilseeds Team, Geneva.

Noble Group Research (2012c): Crush margin calculation of the Grain and Oilseeds Team, Geneva.

Nussbaum, D. (2011): CEO WWF-UK, Speech on the B4E Climate Summit, London.

O'Farrell, I. (2012): Interview with Ion O'Farrell at Noble Argentina on 11th April 2012.

Organisation for Economic Co-operation and Development (OECD, 2011): "Price volatility and agricultural markets: Policy responses", Paris, June 2011, http://www.oecd.org/dataoecd/40/34/48152638.pdf, *accessed on April, 24th 2012.*

Organisation for Economic Co-operation and Development (OECD, 2008): "Biofuel Support Policies: An Economic Assessment", Paris, http://www.oecd.org/document/30/0,3746,en_2649_ 33785_41211998_1_1_1_1,00.html, *accessed on April, 24th 2012.*

Organisation for Economic Co-operation and Development (OECD, 2007): "Biofuels May Be Threat To Food Supply", Increase Costs, http://www.enn.com/energy/article/22940, *accessed on April, 24th 2012.*

Oil World (2009): "Annual Report", http://www.oilworld.biz/.

OVID (2012): "Daten und Grafiken", based on ACTI, GTIS, Oil World and OVID data, Verband der oelsaatenverarbeitenden Industrie, http://www.ovid-verband.de/unsere-branche/daten-und-grafiken/, *accessed on April, 24th 2012.*

Oxfam International (2011): "Land and Power, The growing scandal surrounding the new wave of investments in land", http://www.oxfam.org/en/grow/policy/land-and-power, , *accessed on April, 24th 2012.*

Oxford Learners Dictionary (2012): Online English-English Dicrtionary, Oxford Learners Dictionary, http://oaadonline.oxfordlearnersdictionaries.com/, *accessed on April, 24th 2012.*

Parry, M.L. et al. (2009): "Assessing the Costs of Adapting to Climate Change: a Review of the UNFCCC and Other Recent Estimates", London, Imperial College, Grantham Institute for Climate Change.

Passerini, G. (2011): Interview with Gustavo Passerini, Origination Manager of Noble Argentina in Rosario, Argentina on December, 7th 2011.

Pazouki, Mohammad et al. (2010): "Esterification of Free Fatty Acids by Rhizopus oryzae as Cell-Catalyzed from Used Cooking Oil for Biodiesel Production", World Applied Sciences Journal 8 (6): 719-724, 2010 , http://www.idosi.org/wasj/wasj8(6)10/10.pdf, *accessed on April, 24th 2012.*

Pelkmans, L. et al. (2008): "Inventory of biofuel policy measures and their impact on the market", Report of ELOBIO subtasks 2.1-2.2, http://www.elobio.eu/fileadmin/elobio/user/docs/Elobio_D2_1_PolicyInventory.pdf, *accessed on April, 24th 2012.*

Pradhan, A. et al. (2011): "Energy life cycle assessment if soybean biodiesel revisited", American Society of Agricultural and Biological Engineers ISSN 2151-0032, http://www.usda.gov/oce/reports/energy/EnergyLifeCycleSoybeanBiodiesel6-11.pdf, *accessed on April, 24th 2012.*

Pradhan, A. et al. (2009): "Energy life-cycle assessment of soybean biodiesel", Agricultural Economic Report No. 845, http://www.usda.gov/oce/reports/energy/ELCAofSoybeanBiodiesel91409.pdf, *accessed on April, 24th 2012.*

Pradhan, A. et al. (2008): "The energy balance of soybean oil biodiesel production: A review of past studies", Trans ASABE 51(1): 185-194, http://naldc.nal.usda.gov/download/19610/PDF,

Ragwitz, M. et al. (2005): "Analysis of the renewable energy's evolution up to 2020, FORRES 2020, Fraunhofer IRB Verlag, ISBN 3-8167-6893-8, http://www.eeg.tuwien.ac.at/eeg.tuwien.ac.at_pages/research/downloads/PR_30_FORRES_summary.pdf,

Rain Forest Alliance (2012): Website http://www.rainforest-alliance.org/work/agriculture, *accessed online on February, 11th 2012.*

Renewable Policy Network for the 21st Century (REN21, 2011): "Renewables - Global Status Report", various issues, Paris, 2011.

Robles, M. et al. (2009): "When Speculation Matters" in: IFPRI Issue Brief, February 2009, http://www.ifpri.org/publication/when-speculation-matters, *accessed online on February, 11th 2012.*

Rothkopf, G. (2007): A Blueprint for Green Energy in the Americas", Inter American Development Bank, http://www.scribd.com/doc/35191122/A-Blueprint-for-Green-Energy-in-the-Americas-Prepared-for-the-Inter-American-Development-Bank-by-Garten-Rothkopf, *accessed on February, 17th 2012.*

Roundtable on Responsible Soy (2012): Organisation website, http://www.responsiblesoy.org/, *accessed online on February, 11th 2012.*

Roundtable on Sustainable Biofuels (RSB, 2012): Organisation website, http://rsb.epfl.ch/, *accessed online on February, 11th 2012.*

Roundtable on Sustainable Palm Oil (RSPO, 2012): Organisation website, http://www.rspo.org/, *accessed online on February, 11th 2012.*

Runge, C. & Johnson, R. (2008): "The browning of biofuels – environment and food security at risk", Woodrow Wilson International Center for Scholars, Science, Technology, America and the Global Economy, http://www.agritrade.org/documents/Biofuels_Subs_Web_Final.pdf, *accessed online on February, 11th 2012.*

Sanstandards (2012): Organisation Website, http://sanstandards.org/sitio, *accessed online on February, 11th 2012.*

Schmitz, P. M. and Moleva, P. (2011): „Die Weltagrar- und Finanzmärkte: Spekulation, agrarische Rohstoffe ud Hunger - Irrtümer Und Notwendige Klärung" in: Deutsche Landwirtschaftsgesellschaft (Dlg E.V.), Welternährung – Welche Verantwortung Hat Europa? Dlg-Wintertagung 2012, Band 106, Frankfurt Am Main, S. 73-81, http://www.ufop.de/downloads/UFOP_0966_Vorstudie.pdf, *accessed online on February, 11th 2012.*

Schnepf, R. (2006): "European Union Biofuels Policy and Agriculture: CRS Report for Congress", http://www.usembassy.it/pdf/other/RS22404.pdf, *accessed online on February, 11th 2012.*

Searchinger, T. et al. (2008): "Use of U.S. croplands for biofuels increases greenhouse gases through emissions from land-use change", Science, http://www.sciencemag.org/content/319/5867/1238.full, *accessed online on February, 11th 2012.*

Shapouri, H., et al. (2002): "The energy balance of corn ethanol: An update", Washington, D.C.: USDA, Office of the Chief Economist, Office of Energy Policy and New Uses. http://www.usda.gov/oce/reports/energy/aer-814.pdf, *accessed online on February, 11th 2012.*

Sheehan, J. et al. (1998): "Life-cycle inventory of biodiesel and petroleum diesel for use in an urban bus", NREL/SR-580-24089, Golden, Colo.: National Renewable Energy Laboratory, http://www.usda.gov/oce/reports/energy/aer-814.pdf, *accessed online on February, 11th 2012.*

Simantob, M. (2011): Interview with Marcos Simantob, Senior Trader and Director of Grains & Oilseeds Brazil in Sao Paulo, Brazil on December, 16th 2011.

Smeets et al., (2004): "A quickscan of global bio-energy potentials to 2050", http://www.bioenergytrade .org/downloads/smeetsglobalquickscan2050.pdf, *accessed online on February, 11th 2012.*

Soroko, M. (2012): Interview with Mikhail Soroko, Trading Manager for Black Sea Region at Noble Group, on February, 14th 2012.

Stakeholder Forum (2011): "A pocket Guide to Sustainable Development Governance" http://www.stakeholderforum.org/fileadmin/files/sdgpocketguideFINAL-no%20crop%20marks.pdf, *accessed online on February, 11th 2012.*

Steer, A. (2011): "Special Envoy for Climate Change", The World Bank, Business for the Environement Conference London, B4E CLimate Summit 2011.

Stern, N. (2010), "Stern Review: The Economics of Climate Change", http://news.bbc.co.uk /2/shared/bsp/hi/pdfs/30_10_06_exec_sum.pdf, *accessed online on February, 11th 2012.*

Stern, N. (2008): "Key Elements of a Global Deal on Climate Change", London, London School of Economics and Political Science, http://rrpec.org/documents/ClimateChangeLSE.pdf, *accessed online on February, 11th 2012.*

Stevens, P. (2008): "The Coming Oil Supply Crunch", London, Chatham House, http://www.scribd.com/doc/6900411/The-Coming-Oil-Supply-Crunch, *accessed online on February, 11th 2012.*

Taheripour, F. (2010): "Biofuels and their by-products - Global economic and environmental implications", Biomass and Bioenergy 34, 278 – 289, http://www.agecon.purdue.edu/papers/biofuels/Taheripour_et_al_Biomass_Bioenergy.pdf, *accessed online on February, 11th 2012.*

Teulon, H. (2011): "Life Cycle Analysis and Ecodesign", 12 hour module at HEC Paris within the Master of Science in Sustainable Development 2010-2011, Paris.

Timilsina, G. and Shrestha, A. (2010): "Biofuels - Markets, Targets and Impacts", Policy Research Working Paper 5364, World Bank, Develop Research Group, http://www.scribd.com/ doc/35803671/Biofuels-Markets-Targets-and-Impacts, *accessed on January, 10th 2012.*

Tomei, J. and Upham, P. (2009): "Argentine soy-based biodiesel: An introduction to production and impacts", Energy Policy 37 (10), 3890 – 3898, http://www.scribd.com/doc/14787586/Argentinean-soy-based-biodiesel-an-introduction-to-production-and-impacts, *accessed on January, 10th 2012.*

Trostle, R. (2008): "Global Agricultural Supply and Demand: Factors Contributing to the Recent Increase in Food Commodity Prices", Washington DC, http://www.growthforce.orgwww.growthenergy.org/images/reports/USDA_Global_Agricultural_Supply_and_Demand.pdf, *accessed on January, 10th 2012.*

United Nations Department of Economic and Social Affairs (2006): "World Population Prospects: the 2006 Revision", Economic and Social Affairs, New York, http://www.un.org/esa/population/publications/wpp2006/WPP2006_Highlights_rev.pdf, *accessed on January, 10th 2012.*

United Nations Environmental Programme (UNEP, 2009): "Toward sustainable production and use of resources: Assessing biofuels" http://www.unep.fr/scp/rpanel/pdf/assessing_biofuels_full_ report.pdf, *accessed on January, 10th 2012.*

United Nations High Level Task Force on the Global Food Crisis (2008): "Comprehensive Framework for Action", New York, http://www.ifad.org/operations/food/documents/cfa/cfa_ draft.pdf, *accessed on January, 10th 2012.*

US Department of Agriculture, Foreign Agricultural Service (USDA, 2009a): "Argentina Biofuels Annual", http://gain.fas.usda.gov/Recent%20GAIN%20Publications/General%20Report_ Buenos%20Aires_Argentina_6-19-2009.pdf, *accessed on January, 10th 2012.*

US Department of Agriculture, Foreign Agricultural Service (USDA, 2009b): "Honduras Biofuels Annual", http://gain.fas.usda.gov/Recent%20GAIN%20Publications/General%20Report_ Tegucigalpa_Honduras_5-28-2009.pdf, *accessed on January, 10th 2012.*

US Department of Agriculture, Foreign Agricultural Service (USDA, 2009c): "China Biofuels Annual", http://gain.fas.usda.gov/Recent%20GAIN%20Publications/BIOFUELS%20ANNUAL_ Beijing_China%20-%20Peoples%20Republic%20of_2009-7-17.doc.pdf, *accessed on January, 10th 2012.*

US Department of Agriculture, Foreign Agricultural Service (USDA, 2009d): "Colombia Biofuels Annual", http://gain.fas.usda.gov/Recent%20GAIN%20Publications/General%20Report_Bogota_ Colombia_5-8-2009.pdf, *accessed on January, 10th 2012.*

US Department of Agriculture, Foreign Agricultural Service (USDA, 2010a): "A USDA regional roadmap to meeting the biofuels goals of the Renewable Fuels Standard by 2022", Washington DC, http://www.usda.gov/documents/USDA_Biofuels_Report_6232010.pdf, *accessed on January, 10th 2012.*

US Department of Agriculture, Foreign Agricultural Service (USDA, 2010b): "Brazil - Biofuels Annual", Sao Paulo, http://gain.fas.usda.gov/Recent%20GAIN%20Publications/Biofuels%20 Annual_Sao%20Paulo%20ATO_Brazil_8-11-2010.pdf, *accessed on January, 10th 2012.*

US Department of Agriculture, Foreign Agricultural Service (USDA, 2010c): "EU27 - Biofuels Annual", The Hague, http://gain.fas.usda.gov/Recent%20GAIN%20Publications/Biofuels%20 Annual_The%20Hague_EU-27_6-11-2010.pdf, *accessed on January, 10th 2012.*

US Department of Agriculture, Foreign Agricultural Service (USDA, 2011a): "Argentina - Biofuels Annual", Buenos Aires, http://gain.fas.usda.gov/Recent%20GAIN%20Publications/Biofuels%20 Annual_Buenos%20Aires_Argentina_7-8-2011.pdf, *accessed on January, 10th 2012.*

US Department of Agriculture, Foreign Agricultural Service (USDA, 2011b): "Brazil - Biofuels Annual", Sao Paulo, http://gain.fas.usda.gov/Recent%20GAIN%20Publications/Biofuels%20 Annual_Sao%20Paulo%20ATO_Brazil_7-27-2011.pdf, *accessed on January, 10th 2012.*

US Department of Agriculture, Foreign Agricultural Service (USDA, 2011c): "Global Agricultural Trade System (GATS)", Washington DC, http://www.fas.usda.gov/gats/, *accessed on January, 10th 2012.*

US Department of Agriculture, Foreign Agricultural Service (USDA, 2011d): "Indonesia - Biofuels Annual", Jakarta, http://gain.fas.usda.gov/Recent%20GAIN%20Publications/Biofuels%20Annual_Jakarta_In donesia_8-19-2011.pdf, *accessed on January, 10th 2012.*

US Department of Agriculture, Foreign Agricultural Service (USDA, 2011e): „Italian Biofuels 2011", Rome, http://gain.fas.usda.gov/Recent%20GAIN%20Publications/Italian%20Biofuels%202011_ Rome_Italy_12-29-2011.pdf, *accessed on January, 10th 2012.*

US Department of Agriculture, Foreign Agricultural Service (USDA, 2011f): "Malaysia - Biofuels Annual", Kuala Lumpur, http://gain.fas.usda.gov/Recent%20GAIN%20Publications/Biofuels%20 Annual_Kuala%20Lumpur_Malaysia_8-3-2011.pdf, *accessed on January, 10th 2012.*

US Department of Agriculture, Foreign Agricultural Service (USDA, 2011g): "FAQs on Biofuel Sustainability Certification in Germany", http://gain.fas.usda.gov/Recent%20GAIN%20 Publications/FAQs%20on%20Biofuel%20Sustainability%20Certification%20in%20Germany_Berlin_Germany_04-01-2011.pdf, *accessed on January, 10th 2012.*

US Department of Agriculture, Foreign Agricultural Service (USDA, 2011h): "China – Peoples Report of - Biofuels Annual – 2011 Annual Report", http://gain.fas.usda.gov/Recent%20GAIN%20Publications/Biofuels%20Annual_Beijing_China%20-%20Peoples%20Republic%20of_7-21-2011.pdf, *accessed on January, 10th 2012.*

US Department of Agriculture, Foreign Agricultural Service (USDA, 2012): "Adoption of Genetically Engineered Crops in the U.S.", Economic Research Service, http://www.ers.usda.gov/Data/BiotechCrops/, *accessed on January, 10th 2012.*

US Department of Energy – Energy Efficiency and Renewable Energy (EERE, 2012): "Fuel Flex Vehicle", http://www.fueleconomy.gov/feg/flextech.shtml, *accessed on April, 20th 2012.*

US Energy Information Administration (EIA, 2010): United States energy profile. Washington, D.C.: U.S. Department of Energy, http://tonto.eia.doe.gov/country/country_energy_data.cfm?fips=US, *accessed on January, 10th 2012.*

US Energy Information Administration (EIA, 2011): Monthly Energy Review. Washington DC, http://205.254.135.7/totalenergy/data/monthly/, *accessed on December, January, February, March, April 2012.*

US Environmental Protection Agency (EPA, 2004) "National emission standards for hazardous air pollutants - Solvent extraction for vegetable oil production", 40 CFR Part 63. Washington, D.C., http://www.epa.gov/ttn/atw/vegoil/fr01se04.pdf, *accessed on December, 9th 2012.*

Van Gerpen, J. (2005): "Biodiesel processing and production", http://users.rowan.edu/~marchese/froshclinic2-07/vangerpen.pdf, *accessed on January, 04th 2012.*

Vasquez, G. (2012): Interview with Gustavo de Vasquez, Executive and former head of Biodiesel Production of Archer Daniel Midlands on February, 14th 2012.

Verdonk, M. et al. (2007): "Governance of the emerging bio-energy markets", Utrecht, http://www.lowcvp.org.uk/assets/reports/governance%20biomarkets%20verdonk%20faaij%20dieperink%20article%20energy%20policy.pdf, *accessed on April, 20th 2012.*

Von Braun, J. (2007): "The World Food Situation: New Driving Forces and Required Actions", Washington DC, IFPRI, http://www.ifpri.org/sites/default/files/pubs/pubs/fpr/pr18.pdf, *accessed on on February, 14th 2012.*

Wang, M. Q., and H. S. Huang (1999): "A full fuel-cycle analysis of energy and emissions impacts of transportation fuels produced from natural gas", ANL/ESD-40, Argonne National Laboratory, www.transportation.anl.gov/pdfs/TA/13.pdf, *accessed on on February, 14th 2012.*

WDPA (2012): Definition of protected Areas, http://www.wdpa.org/ and recently http://protectedplanet.net/, *accessed on April, 20th 2012.*

Wiesenthal, T. et al. (2009): "Biofuel support policies in Europe: Lessons learnt for the long way ahead", Renewable and Sustainable Energy Reviews 13: 789–800, http://dx.doi.org/10.1016/j.rser.2008.01.011, *accessed on April, 12th 2011.*

World Bank (2008a): "World Development Report 2008", Washington DC, http://web.worldbank.org/WBSITE/EXTERNAL/NEWS/0,,contentMDK:21513382~pagePK:64257043~piPK:437376~theSitePK:4607,00.html, *accessed on December, 11th 2011.*

World Bank (2008b): "Double Jeopardy: Responding to High Food and Fuel Prices", Working paper for G8 Hokkaido-Toyako summit, 2 July, Washington DC, http://web.worldbank.org/WBSITE/EXTERNAL/NEWS/0,,contentMDK:21827681~pagePK:64257043~piPK:437376~theSitePK:4607,00.html, *accessed on April, 21st 2012.*

World Bank (2010): Research database - Biofuel Briefing, , http://siteresources.worldbank.org/EXTESC/Resources/Biofuel_brief_Web_version.pdf, *accessed on April, 21st 2012.*

World Food Programme (2009): "Climate Change and Hunger - Responding to the Challenge", Published in 2009 by the World Food Programme, http://documents.wfp.org/stellent/groups/public/ documents/newsroom/wfp212536.pdf, *accessed on April, 21st 2012.*

World Research Institute (WRI, 2009): "Climate analysis indicators tools (CAIT)" http://cait.wri.org/downloads/CAIT_7.0_COP15.pdf, "Biofuels and the time Value of Carbon: Recommendations for GHG Accounting Protocols", Working Paper, http://pdf.wri.org/working_papers/time_value_of_carbon.pdf, *accessed on April, 21st 2012.*

World Research Institute (WRI, 2012):" Carbon Dioxide Capture and Storage (CCS)", http://www.wri.org/project/carbon-dioxide-capture-storage, *accessed on April, 21st 2012.*

World Trade Organisation (WTO, 2001): "Market access for non-agricultural products", The Doha Mandate from the Doha Ministerial Declaration, November 2001, https://www.wto.org/english/tratop_e/markacc_e/markacc_dohamandate_e.htm, *accessed on April, 21st 2012.*

Worldwatch Institute (2007): Biofuels for transport: global potential and implications for sustainable energy and agriculture, Earthscan, London p. 148

Worldwatch Institute (2009): Smart Choices for biofuels, http://www.worldwatch.org/files/pdf/biofuels.pdf

WWF (2011): The Energy Report, Wild Wonders of Europe, Inaki Relanzon http://assets.panda.org/downloads/101223_energy_report_final_print_2.pdf, *accessed on April, 21st 2012.*

WWF (2003): "Sustainability Assessment of export-led growth in Soy Production In Brazil", http://vsites.unb.br/face/eco/jmn/textos/SoyaEnvironment.pdf, *accessed on April, 21st 2012.*

WWF International (2011): UN Estimates: Samantha Smith, Lead, Global Climate & Energy Initiative, speech in Jakarta B4E Climate Summit 2011

Zylla, J. (2012): Interview with Joerg Zylla, Director and Global Head of Vegetable Oil at Bunge on 10th January 2012.

Words: 30,920

References: 210

Appendix

Noble Group company presentation

- Bulk global commodity supply chain manager, operating across a multitude of products, locations and customers
- Resilient and diversified business model underpinned by risk management expertise providing stable Group margins
- Global operating platform emphasising seaborne trade with origination in low cost countries, delivery to high growth markets
- Generated revenue of US$56.7 billion and EBITDA of US$1.2 billion in FY 2010
- Headquartered in HK and listed on Singapore Stock Exchange (N21) with market capitalisation of US$11.3 billion as of April 8, 2011
- Investment grade ratings from S&P, Moody's and Fitch
- Net book value per share compound growth of 27% since 1997 listing
- Excluding Oil, Gas & Power, over 85% of tonnage delivered to emerging markets

Noble Group is a Fortune 500 company with a well established worldwide presence

40 Countries, 79 Nationalities, 150 Officies, 61 Assets, 11000+ Employees, #139 in the Fortune 500 Ranking

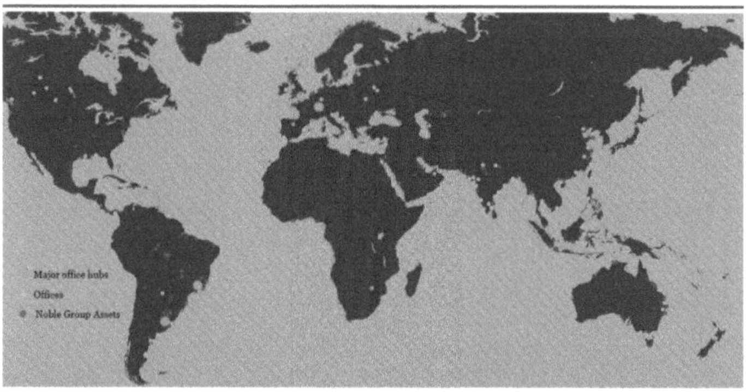

noble group

Almost 50% of the Noble operating income is generated through the Agriculture business

Percentage of operating income from supply chain*

Agriculture	Energy	Metals, Minerals & Ores	Logistics
47%	33%	13%	7%
Transports and processes agricultural commodities such as wheat, soybeans and sugar in key locations	Mines and markets energy products such as coal and coke. Developing an integrated oil, gas & power platform	Manages sourcing and supply of raw materials that drive industry: principally iron ore and aluminium	Offers vessel chartering and the storage and distribution necessary to bring our own, and third party, products to market

A diversified business portfolio helps to create a stable and resilient income overcoming volatile individual product margins

Agriculture segment key 2010 financials

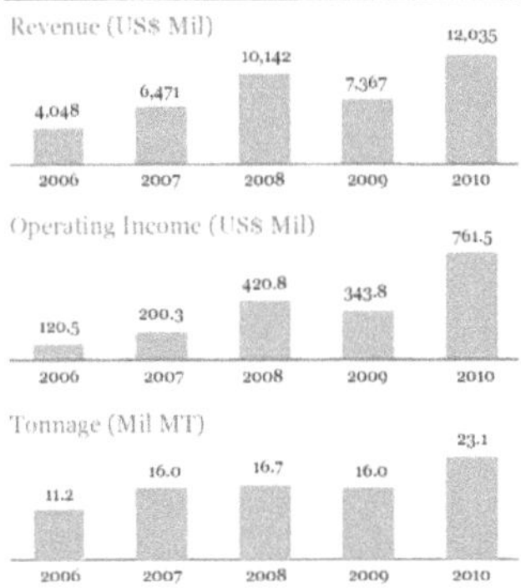

2010 Grain and Oilseeds key developments

- Inauguration of Timbues crushing plant
- Inauguration of Santos export terminal
- Start up of Ming Fa (China), vegetable oil and pal oil refinery
- Start up of International Rice business in Geneva
- Acquisition of Mariupol (Ukraine) sunglower seed crushing plan

2011 Grain and Oilseeds Strategy

- Start the construction of Rondonopolis crushing plant; 2 country elevators in Mato Grosso and 2 country stations in Argentina
- Set up of strong centralized international trading unit in Geneva
- Set up origination and export structure in Australia
- Acquisition of 1 additional greenfield palm plantations in Indonesia / Papua NG

Noble aims at generating profit through multiple points along the entire supply chain

Example of Noble's Pipeline Strategy: Soybean Transaction From Argentina to China

Production	Purchase	Warehouse	River port grain terminal	Logistics & shipping	Processing	Customers
Fertilisers sold to farmers from Noble JV in Brazil	Beans bought directly from farmers in Brazil, Argentina, Paraguay, Uruguay	Products stored in our storage facilities in Brazil, Argentina, Paraguay, Uruguay	Products loaded on vessels at Noble ports in Argentina and Brazil	Logistics and shipping provided by Noble Chartering	Beans are crushed and refined in our integrated plant in China	Sell final products to end-users worldwide

Profit generated through multiple profit points along entire supply chain

Fertiliser sales	Farmer financing	Storage fees	Elevation fees	Freight Income	Crushing and refining margin	Soybean meal (feedstock) & oil (consumer use)
			Or crush at Noble Timbúes plant	Oil to Asia	Meal to Europe	

Noble is present in most of the major trading routes for Grains & Wheat and Soybeans & Oilseeds

Grains & Wheat Major Routes

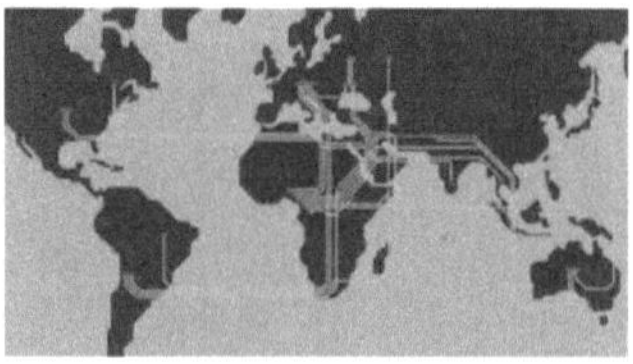

Soybeans & Oilseeds Major Routes

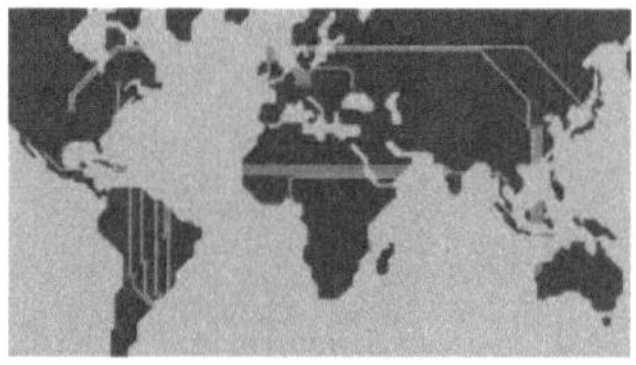

Global operating platform emphasising seaborne trade with origination in low cost countries and delivery to high growth markets

Noble Grain & Oilseeds originated last year more than 17 million metric tons mainly from South America

Grain & Oilseeds origination by product ('000 MT)

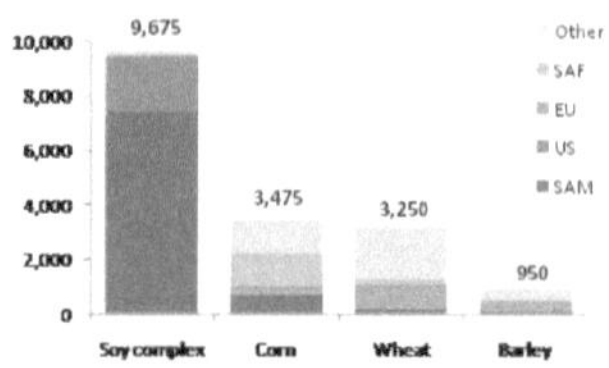

Grain & Oilseeds origination by geography ('000 MT)

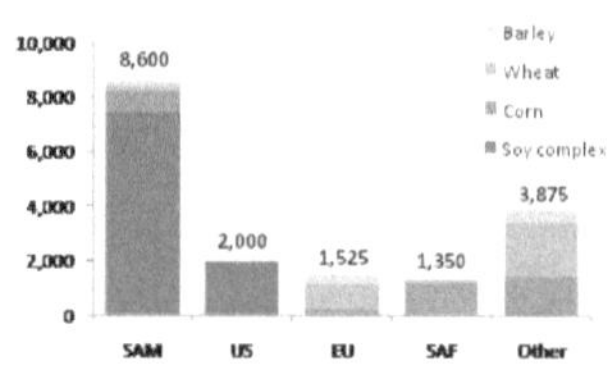

	Soy complex	Corn	Wheat	Barley	Total
	9,675	3,475	3,250	950	17,350
SAM	7,500	750	200	150	8,600
US	2,000				2,000
EU		275	900	350	1,525
SAF		1,200	150		1,350
Other	175	1,250	2,000	450	3,875

noble group

Strategic networks of ports and terminals play a key role in the seamless flow of agri commodity through our global pipelines

Noble Major Ports and Terminals (operating and under-construction)

1. **Argentina:**
   - Timbues grain terminal (5 myn annual through-put)
   - Lima grain terminal (1 myn annual through-put)
2. **Uruguay:** Nueva Palmira grain terminal, equity stake (750k annual through-put)
3. **Brazil:** Santos dry bulk export terminal (2.5myn annual through-put)
4. **Netherlands:** Rotterdam edible/industrial oil terminal, (550k cubic metres storage capacity)
5. **Paraguay:** Pacu Cua river port terminal (400k MT annual through-put)
6. **Russian Ports:** various through-put agreements (725k annual through-put)
7. **Ukraine Ports:** various throughput agreements (800k annual through-put)

Noble Major Processing Facilities (operating and under-construction)

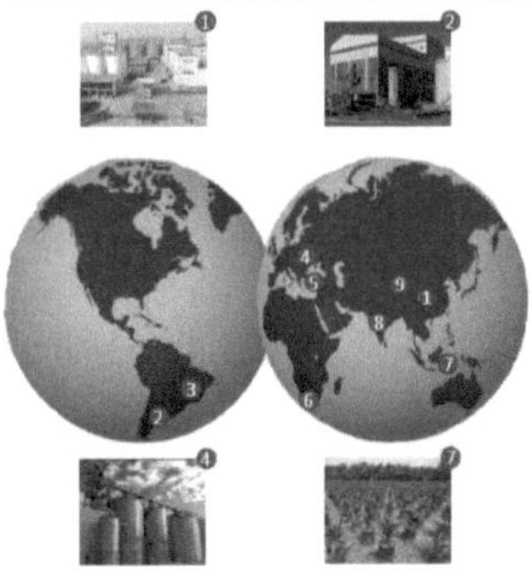

1. **China** 4 Crush Plants approximately 14k mt per day
2. **Argentina** 1 Crush Plant approximately 9k mt per day
3. **Brazil** 1 Crush Plant 4k mt per day
4. **Ukraine** Sunseed crush plant 2k mt per day
5. **Turkey** toll crush facilities 500 mt per day
6. **South African** multi seed crush plant 2k mt per day
7. **Indonesian** Palm Crush (180k annual capacity) and palm plantation land bank (70,000Ha)
8. **India** Kandla Multi-Oil refinery 1k CPO + 400 mt Soft oils per day
9. **China** Taixing Multi-Oil refinery 450mt Olein + 300mt Soft oils per day

Noble Major Storage Facilities (operating and under-construction)

1. **Brazil:** interior warehouses/silos (275k MT capacity)
2. **Argentina:** interior warehouses (80k MT capacity)
3. **Jordan:** silo/warehouse complex (70k MT capacity)
4. **Saudi Arabia:** grain warehouse (150k MT capacity)
5. **China:** grain storage terminal (375k MT capacity)
6. **Ukraine:** interior warehouse (100k MT capacity)
7. **China:** oil storage facilities (200k MT capacity)

Acknowledgements

First of all, I would like to thank the *HEC School of Management* and its professors providing a profound basis and knowledge about sustainability and especially *my tutor Manuel Menuet*, for the guidance, efforts and support during the last semester, which made the realisation of the book possible. Equally, I would like to express my thanks to the *Fundação Getulio Fargas* and its professors that have been teaching within the *MBA in Sustainability* for their time and advices.

Additionally, I would like to thank *Gustavo de Vasquez* from *ADM* for his time and commitment as well as his guidance in relation to *Biodiesel*. Furthermore, I would like to deeply thank my tutor, *Wouter Loeff*, my manager Laurent Guerrero and my mentor and friend, Yenal Ghori, all working at *Noble Resources* for his good advices and contacts as well as the provided knowledge in the field of Supply Chain Management. Moreover, I would like to thank *Gabriela Montenegro*, working for Noble Argentina and head of quality and environment, for her recommendations and fruitful discussions in the field of certification and sustain. Besides, I would like to thank two friends, *Jan Kabath and Felix Heinrich* the former working as Trader at *Louis Dreyfus* responsible for European Vegetable Oil and Biodiesel and the latter working as Research Analyst at *Pareto Securities* Investment Bank, who both supported my work and gave complementary advises as well as personal encouragement.

In addition, I would like to say thank you to my friends, *Amina Adewusi, Chioma Oganya, Emma Cawood, Mike Wedderburn* and *Barnaby Harris*, who always had time and the necessary patience to listen to me and conveying me another point of view about my book. Finally, my honest thanks and gratitude are directed to my family, my parents *Thomas* and *Sonja* as well as my sister *Sarah*, who always supported me and my ideas, providing me a very strong hold and security, which was the basis of my purposes and realised projects.